Lecture Notes in Economics and Mathematical Systems

Operations Research, Computer Science, Social Science

Edited by M. Beckmann, Providence, G. Goos, Karlsruhe, and H. P. Künzi, Zürich

79

A. Berman

Geschäftsbibliothek
Springer-Yerlag, Berlin

Cones, Matrices and Mathematical Programming

Springer-Verlag
Berlin · Heidelberg · New York 1973

Dr. Abraham Berman
Centre de Recherches Mathématiques
Montréal and Department of Mathematics
Israel Institute of Technology
Haifa, Israel

AMS Subject Classifications (1970): Primary   15-02, 15 A 39, 15 A 48, 52-02, 90-02, 90 C 05, 90 C 20, 90 C 25, 90 C 30,
                                    Secondary   15 A 09, 15 A 42, 15 A 45, 15 A 60, 15 A 63, 52 A 25, 52 A 40, 90 C 25

ISBN -13:978-3-540-06123-6   e- ISBN -13:978-3-642-80730-5
DOI: 10.1007/978-3-642-80730-5

PREFACE

This monograph is a revised set of notes on recent applications of the theory of cones, arising from lectures I gave during my stay at the Centre de recherches mathématiques in Montreal. It consists of three chapters. The first describes the basic theory. The second is devoted to applications to mathematical programming and the third to matrix theory.

The second and third chapters are independent. Natural links between them, such as mathematical programming over matrix cones, are only mentioned in passing.

The choice of applications described in this paper is a reflection of my personal interests, for examples, the complementarity problem and iterative methods for singular systems. The paper definitely does not contain all the applications which fit its title. The same remark holds for the list of references.

Proofs are omitted or sketched briefly unless they are very simple. However, I have tried to include proofs of results which are not widely available, e.g. results in preprints or reports, and proofs, based on the theory of cones, of classical theorems.

This monograph benefited from helpful discussions with professors Abrams, Barker, Cottle, Fan, Plemmons, Schneider, Taussky and Varga. I am especially grateful to Professor A. Ben-Israel, whose guidance and cooperation during my study at the Israel Institute of Technology and, at Northwestern University, have stimulated my interest in the material presented here. Finally, I am very grateful to Professor L. Lecam, director of the Centre de recherches mathématiques for encouraging me to write these notes, and to Madame Janine Viau of the Centre for her patience and good will in typing the manuscript.

Abraham Berman
Montreal, August 1972

CONTENTS

CONVEX CONES and LINEAR INEQUALITIES

In this chapter we describe basic results in the theory of cones. Having the applications in mind the discussion is restricted to finite dimensional spaces. Work in general vector spaces is mentioned in the references.

Convex cones are defined and studied following preliminaries on separation theorems. Linear inequalities and theorems of the alternative are studied in the last two sections of this chapter.

1. Separation Theorems

The study of linear inequalities is essentially based on separation of convex sets.

For a survey of the theory of separation the reader is referred to some of the basic references: Eggleston [1], Klee [2] and Rockafellar [1] for the finite dimensional case, Bourbaki [1], Day [1], Fan [1] and Schaefer [1] for normed or general topological vector spaces and Deutsch and Maserick [1], where applications to approximation theory are given.

Here we collect several separation theorems in C^n, the n-dimensional complex space, that will be used in the sequel.

A set K in C^n is __convex__ if it is closed under convex combinations, i.e.,

$$(1.1) \qquad \alpha K + (1-\alpha)K \subseteq K \quad \text{for } 0 \leq \alpha \leq 1$$

K is a _cone_ if it is closed under nonnegative scalar multiplication, i.e.,

(1.2) $\alpha K \subseteq K$ if $\alpha \geq 0$.

K is a _convex cone_ if it satisfies (1.1) and (1.2), or equivalently, (1.2) and

(1.3) $K + K \subseteq K$.

A _hyperplane_ H in C^n is a set of the form

(1.4) $H = \left\{ x \in C^n; \ \text{Re} \ (u,x) = \alpha \right\}$

where u is a non-zero vector in C^n, α is a real number and (,) denotes the inner product in C^n.

A _manifold_ (or flat or linear variety) in C^n is a set of the form $S + x_0$, where S is a linear subspace and x_0 is a fixed point in C^n. H, given in (1.4) is a manifold, where S is a subspace of dimension n-1 and Re $(u,x_0) = \alpha$.

Let H be the hyperplane given by (1.4). Then

(1.5) $\left\{ x \in C^n; \ \text{Re} \ (u,x) \leq \alpha \right\}$

and

$$(1.5)' \qquad \left\{ x \in C^n; \ \text{Re} \ (u,x) \geqq \alpha \right\}$$

are closed sets, called the <u>closed half spaces</u> determined by H, and the sets

$$(1.6) \qquad \left\{ x \in C^n; \ \text{Re} \ (u,x) < \alpha \right\}$$

and

$$(1.6)' \qquad \left\{ x \in C^n; \ \text{Re} \ (u,x) > \alpha \right\}$$

are open sets, the <u>open half spaces</u> determined by H.

A set S is said to <u>lie on one side</u> of a hyperplane H if S is contained in one of the two closed half spaces determined by H. S is said to <u>strictly lie</u> on one side of H, if it is contained in one of the two open half spaces determined by H. Notice that if an open set is contained in (1.5) [(1.5)'] then it lies in (1.6) [(1.6)'].

A hyperplane H is said to <u>separate</u> two non empty sets S and T, if S is contained in one of the two closed half spaces determined by H and if T is contained in the other open half space.

A separation theorem deals with the existence of a separation of two given objects. The following separation theorems cover the cases where these objects are a closed convex set S and a point outside S, an open convex set S and a manifold not meeting S, and two disjoint convex sets, one of them open.

The proofs are modifications to the complex case of special cases of results in the references mentioned above, and will be omitted.

Theorem 1.1 Let $S \subset C^n$ be a closed convex set, $b \in C^n$, $b \notin S$. Then there is a hyperplane, given by (1.4) such that

(1.7) $\mathrm{Re}\ (u,x) \geqq \alpha\quad \text{for}\quad x \in S$

(1.8) $\mathrm{Re}\ (u,b) < \alpha$

Corollary 1.1. Let K be a closed convex cone in C^n, $b \in C^n$, $b \notin K$. Then there is a hyperplane passing through the origin and separating b and S, i.e., there exists $0 \neq u \in C^n$ such that

$$\mathrm{Re}\ (u,x) \geqq 0\quad \text{for}\quad x \in K$$

and $\mathrm{Re}\ (u,b) < 0$.

Theorem 1.2 (The geometric version of the Hahn-Banach Theorem, or Mazur's Theorem). Let S be an open convex set in C^n, and let M be a manifold in C^n such that $M \cap S = \emptyset$. Then there exists a hyperplane H, containing M, such that S lies (strictly, since S is open) on one side of H.

Corollary 1.2. Let S_1 be an open convex set and S_2 a convex set in C^n such that $S_1 \cap S_2 = \emptyset$. Then there exists a hyperplane H separating S_1 and S_2.

Corollary 1.3. Let int K be the interior of a convex cone, S a convex set, $S \cap$ int $K = \emptyset$. Then there exists a hyperplane passing through the origin and separating S and int K.

2. Cones and Duals

Let S be a set in C^n. The set

$$S^* = \left\{ y \in C^n; \ \text{Re} \ (x,y) \geq 0 \ \text{if} \ x \in S \right\}$$

is called the dual of S.

Let L be a subspace of C^n. Then $L^* = L^\perp$, the orthogonal complement of L. In particular $\{C^n\}^* = \{0\}$ and $\{0\}^* = C^n$. (In the real case, i.e., sets in R^n - the n dimensional real vector space, $\{R^n\}^* = \{0\}$ and $\{0\}^* = R^n$.) A result of Gaddum ([1], Theorem 2.1) states that $K \cap K^* = \{0\}$ if and only if K is a real subspace.

The following properties of S^* follow from its definition:

Theorem 2.1 (e.g. Ben-Israel [1]). Let S_1 and S_2 be two non empty sets in C^n. Then

    (a) S_1^* is a closed convex cone.

    (b) $S_1 \subseteq S_2 \Rightarrow S_2^* \subseteq S_1^*$.

    (c) $S \subseteq S^{**}$.

    (d) $S^* = S^{***}$.

    (e) $S^* = \{cl \ S\}^* = \{cone \ S\}^* = \{conv \ S\}^*$.

where {cone S} is the smallest cone containing S, {conv S} is the convex hull of S, i.e., the smallest convex set containing S, and cl denotes closure·

    (f) $S_1^* \cap S_2^* \subseteq (S_1 + S_2)^*$.

        $(S_1 + S_2)^* \subseteq S_1^* \cap S_2^*$ if $0 \in S_1 + S_2$.

    (g) $S_1^* + S_2^* \subseteq (S_1 \cap S_2)^*$.

Let $S_j \subseteq C^{n(j)}$, $j = (1,\ldots,k)$, and let $S = S_1 \times S_2 \times \ldots \times S_k \subset C^{\Sigma n(j)}$

be their cartesian product. Then

$$\text{(h)} \quad S^* = S_1^* \times S_2^* \times \ldots \times S_k^*.$$

The following theorem is basic in the theory of linear inequalities, described in the next section. It is derived from the first separation theorem.

Theorem 2.2 (Farkas). Let $S \subset C^n$. Then S is a closed convex cone if and only if $S = S^{**}$.

Proof. If. By Theorem 2.1 (a).

Only if. By Theorem 2.1 (c), it is enough to show that $S^{**} \subset S$. Suppose $x \notin S$. Then by Corollary 1.1 there is a $u \in S^*$ such that $(u,x) < 0$. Thus $x \notin S^{**}$.

Since the closure of a convex cone is a closed convex cone, one gets:

Corollary 2.1 (Ben-Israel [1]). If K is a convex cone, then $cl\ K = K^{**}$.

More can be said now on the relation (g) in Theorem 2.1.

Corollary 2.2. (Ben-Israel [1]). Let K_1 and K_2 be closed convex cones in C^n. Then $cl\ (K_1^* + K_2^*) = (K_1 \cap K_2)^*$.

Proof.
$$\begin{aligned}
(K_1 \cap K_2)^* &= (K_1^{**} \cap K_2^{**})^* \quad \text{(by Theorem 2.2)} \\
&= (K_1^* + K_2^*)^{**} \quad \text{(by Theorem 2.1 (f))} \\
&= cl\ (K_1^* + K_2^*) \quad \text{(by Corollary 2.1 and since the}
\end{aligned}$$
sum of two convex cones is a convex cone)

Remark. Although K_1^* and K_2^* are closed convex cones, their sum does not have to be closed. This is demonstrated by the following (real)example:

Let K_1^* be the cone of vectors in R^3 forming an angle $\leq \frac{\pi}{4}$ with the vector $\begin{pmatrix} 1 \\ 0 \\ 1 \end{pmatrix}$, i.e.,

$$K_1^* = \left\{ \begin{pmatrix} x_1 \\ x_2 \\ x_3 \end{pmatrix}; x_1 \geq 0, \ x_3 \geq 0, \ 2x_1 x_3 \geq x_2^2 \right\}$$

Let K_2^* be the X-axis. Then

$$\begin{pmatrix} k \\ 1 \\ \frac{1}{2k} \end{pmatrix} + \begin{pmatrix} -k \\ 0 \\ 0 \end{pmatrix} = \begin{pmatrix} 0 \\ 1 \\ \frac{1}{2k} \end{pmatrix} \in K_1^* + K_2^*$$

$$\begin{pmatrix} 0 \\ 1 \\ 0 \end{pmatrix} \in cl \ (K_1^* + K_2^*)$$

but

$$\begin{pmatrix} 0 \\ 1 \\ 0 \end{pmatrix} \notin K_1^* + K_2^*.$$

Geometrically, $K_1^* + K_2^*$ consists of the line $x_2 = x_3 = 0$ and of the open half space $x_3 > 0$ and thus is not closed.

Closedness criteria for the sum of closed convex cones will be given in the following section.

Let K be a closed convex cone in C^n. The largest subspace which is contained in K is $K \cap (-K)$. The smallest subspace which contains K is $K - K$. A convex cone K is <u>pointed</u> if $K \cap -(K) = \{0\}$, <u>reproducing</u> if $K-K = C^n$, and <u>solid</u> if int K, the interior of K is not empty.

A pointed closed convex cone $K \subset C^n$ induces a partial order in C^n via $x \leq y$ if and only if $y - x \in K$.

If K is also solid and $u \in$ int K, the norm defined on C^n by $||x||_u = \mathbf{inf}\{\alpha; \ x+\alpha u \in K, \ -x + \alpha u \in K\}$ is monotone, i.e.

$y-x \in K$, $x \in K \Rightarrow ||\ y\ ||_u \geqq ||\ x\ ||_u$, e.g. Householder [1] p 38. The natural (component wise) partial order in R^n, is the one induced by R_+^n, the nonnegative orthant in R^n.

Lemma 2.1 A closed convex cone in C^n is solid if and only if it is reproducing.

Proof. Only if. Let v be an interior point of K. Then for every $x \in C^n$ there is a sufficiently small $\delta > 0$ such that $u = v + \delta x$ is an interior point of K and $x = \frac{u}{\delta} - \frac{v}{\delta}$, i.e., $C^n = \text{int } K - \text{int } K$. Thus K is reproducing.

If. Suppose K is not reproducing. Then $K - K$ is of dimension less than n and K is not solid.

Lemma 2.1 is not true in a Banach space, e.g., Krasnoselskii [1], p. 18.

Theorem 2.3 (Krein and Rutman [1]). Let K be a closed convex cone in C^n. Then K is pointed if and only if K^* is solid.

Proof. $K \cap (-K)$ is a subspace, thus $(K \cap (-K))^\perp = (K \cap (-K))^* = \text{cl}(K^* - K^*)$ (by Corollary 2.2) $= (K^* - K^*)$ (since $K^* - K^*$ is a subspace, thus closed). If K is pointed, then K^* is reproducing, and, by Lemma 2.1, solid. If K is not pointed, then the dimension of $K^* - K^*$ is less than n. Thus K^* is not reproducing and not solid.

Let K be a pointed closed convex cone in C^n. Then int K^* is given, algebraically, by

$$(2.1) \qquad \text{int } K^* = \left\{ y \in K^*;\ \text{Re}(x,y) > 0 \text{ if } 0 \neq x \in K \right\}.$$

Let K be a closed convex cone. Then K is solid if and only if K^* is pointed, in which case:

$$(2.2) \qquad \text{int } K = \left\{ x \in K; \; 0 \neq y \in K^* \Rightarrow \text{Re}(x,y) > 0 \right\}.$$

We shall call a cone K a <u>full</u> cone if it is closed, convex, pointed and solid. A <u>face</u> of a full cone K is a subset of K, which is a pointed closed convex cone F such that $x - y \epsilon K$, $x \epsilon F$, $y \epsilon K \Rightarrow y \epsilon F$. $F \notin \{\{0\}, K\}$ is called <u>proper</u>.

The following is a simple corollary of (2.2).

<u>Corollary 2.3</u>. Let $K_j \subset C^{n(j)}$, $j = 1, \ldots, k$, be convex cones with nonempty interiors and let $K = K_1 \times K_2 \times \ldots \times K_k \subset C^{\Sigma n(j)}$, be their cartesian product. Then

  (a) $K_1 + \text{int } K_1 \subset \text{int } K_1$

  (b) $\text{int } K = \text{int } K_1 \times \text{int } K_2 \times \ldots \times \text{int } K_k$.

A <u>polyhedral cone</u> is a convex cone generated by finitely many vectors, that is, a set of the form $K = BR_+^k$, for $B \subset C^{n \times k}$. The following theorem gives, without proof, basic properties of polyhedral cones, e.g. Klee [1], Rockafellar [1] pp 170-178.

<u>Theorem 2.4</u>. (a) K is a polyhedral cone in C^n if and only if it is the intersection of finitely many closed half spaces, each containing the origin in its boundary: $S = \bigcap\limits_{k=1}^{p} H_{u_k}$, where $H_{u_k} = \{z \in C^n ; \text{Re }(z, u_k) \geq 0\}$.

  (b)   A polyhedral cone is a closed convex cone.

  (c) K is a polyhedral cone if and only if K^* is a polyhedral cone.

  (d)   The sum of polyhedral cones is polyhedral.

  (e)   The cartesian product of polyhedral cones is a polyhedral cone.

R_+^n, R^n and C^n are examples of polyhedral cones. Another example which will be frequently used in the second chapter is

$$T_\alpha = \{z \in C^n ; |\arg z_i| \leq \alpha_i\}$$

where $0 \leqq \alpha = (\alpha_i) \leqq \frac{\pi}{2}e$, e being a vector of ones.

The duals of these cones are:

$$R_+^{n*} = R_+^n$$

$$R^{n*} = \{0\}, \quad C^{n*} = \{0\}$$

$$T_\alpha{}^* = T_{\frac{\pi}{2}e-\alpha}$$

The following result is a generalization of the decomposition of a space into the sum of two orthogonal complementary subspaces.

Theorem 2.5 (Moreau [1]). Let K be a closed convex cone in C^n. Then every point $x \in C^n$ can be represented uniquely as $x = y + z$, where $y \in K$, $z \in -K^*$ and $Re(y,z) = 0$.

Proof. The proof presented here is based on the following lemma.

Lemma 2.2. Let S be a closed convex set in C^n. Then the following are equivalent:

      (a)   The point c is closest to b in S.

      (b)   If $x \in S$, then $Re(c-b, x-c) \geqq 0$.

Proof. e.g. Ben-Israel [1], p 372.

Proof of Theorem 2.5. Let y be the closest point to x in K. Let $z = x-y$. By Lemma 2.2,

(2.3) $$w \in K \Rightarrow Re(y-x, w-y) \geqq 0$$

Take $w = \alpha y$. Then $Re(y-x, (\alpha-1)y) \geqq 0$ for every $\alpha \geqq 0$. Thus

(2.4) $$Re(y-x, y) = 0.$$

i.e. $Re(z,y) = 0$.

Substituting (2.4) in (2.1) one sees that $\text{Re}(-z,w) \geq 0$, i.e. $z \in -K^*$.

To prove the uniqueness, let $x = y + z$, $y \in K$, $z \in -K^*$, $\text{Re}(y,z) \geq 0$. Then $\text{Re}(y-x, w-y) = \text{Re}(y-x,w) = \text{Re}(-z,w) \geq 0$ for every $w \in K$. Thus by Lemma 2.2, y is the closest point in K to x, and is unique since K is convex. By changing the role of K and $-K^*$ it follows that z is the closest point to x in $-K^*$.

Remark. An extension of Moreau's idea is given in Zarantonello [1].

Corollary 2.4 (Haynsworth and Hoffman [1]). A closed convex cone K contains its dual K^*, if and only if, for each vector $x \in C^n$ there exist vectors y and t in K such that

$$(2.5) \qquad\qquad x = y-t, \; \text{Re}(y,t) = 0$$

Proof. Let $K^* \subseteq K$. Decompose $x = y+z$, as in Theorem 2.5. Then $t = -z \in K^*$ and y satisfy (2.5).

Conversely. Assume that there exists a vector $x \in K^*$ such that $x \notin K$. Decompose $x = y-t$, with y and t satisfying (2.5). Since x is not in K, t is different from zero. Thus $0 \leq \text{Re}(t,x) = \text{Re}(t,y-t) = -||t||^2 < 0$. A contradiction.

The convex cone, {cone S}, was defined above, as the smallest convex cone containing the set S. If the convex hull of S, i.e., the smallest convex set containing S, is compact and does not contain the origin, then {cone S} is closed.

Another method of deriving real cones from a (convex real) set is given next.

Let S be a convex set in R^n with $0 \in S$. The associated cone $C(S)$ in R^{n+1} is defined (Ben-Israel, Charnes and Kortanek [3])by

$$C(S) = \left\{ \begin{pmatrix} \lambda x \\ \lambda \end{pmatrix}, \lambda \geq 0, x \in S \right\}.$$

C(S) is a convex cone. Let $\sigma(S)$ denote the maximal closed convex cone contained in S. Then the closure of $C(S)$ is given by $C(S) \cup \left\{ \begin{matrix} \sigma(S) \\ 0 \end{matrix} \right\}$, and its dual by $\{C(S)\}^* = C\{S^*-1\} \cup \left\{ \begin{matrix} S \\ 0 \end{matrix}^* \right\}$ where $S^*-1 = \{y \in R^n; x \in S \Rightarrow (y,x) \geq -1\}$.

This section is concluded with some of the basic references on cones and duals:

Cones in R^n are discussed in Rockafellar [1].

For polyhedral cones, see Klee [1].

Cones in C^n are discussed in Ben-Israel [1] where the term "polar" is used for dual.

Cones in a Banach space are studied in Krasnoselskii [1] and Krein and Rutman [1]. They call a "cone" what we defined as a "full cone". Krasnoselskii used "conjugate" for dual. Krein and Rutman use the term "linear semi group" for what we defined as a convex cone. The theory of cones in topological vector spaces is developed in Fan[1] and Schaefer [1] where "proper" is used for pointed.

3. Linear inequalities over cones

Real linear inequalities can be represented as linear equations over suitably defined convex cones. For example, the system

$$Bu \leq b$$

where $B \in R^{n \times k}$ (B is a real nxk matrix) and $b \in R^m$, can be rewritten as

(3.1) $Ax = b, \; x \in K$

where $A = (B, I)$ and $K = R^k \times R^m_+$.

Complex linear inequalities are systems like (3.1) with complex data. We denote by $C^{m \times n}$ the mxn complex matrices.

Let $A \in C^{m \times n}$, $b \in C^m$ and let K be a closed convex cone in C^n. The system (3.1) is

(i) consistent if there is a vector x satisfying it, or equivalently if $b \in AK$.

(ii) asymptotically consistent if there is a sequence $\{x_k; \; k = 1, 2, \ldots\} \subset K$ such that $\lim Ax_k = b$, or equivalently if $b \in cl \; AK$.

Theorem 3.1 (Ben-Israel [1]). Let $A \in C^{m \times n}$, $b \in C^m$ and let $K \subset C^n$ be a non empty closed convex cone. Then the system (3.1) is asymptotically consistent if and only if

(3.2) $\qquad\qquad A^H y \in K^* \Rightarrow \mathrm{Re}(b,y) \geq 0.$

Proof (R.A. Abrams).

The system (3.1) is asymptotically consistent $\Longleftrightarrow$

$b \in \mathrm{cl}\ AK \Longleftrightarrow$

$b \in (AK)^{**}$ (by Corollary 2.1) $\Longleftrightarrow$

$[y \in (AK)^* \Rightarrow \mathrm{Re}(b,y) \geq 0] \Longleftrightarrow (3.2)$

Theorem 3.2 (Ben Israel [1]). Let A,b and K be as in Theorem 3.1 and let N(A) + K be closed. Then the following are equivalent:

(a) The system (3.1) is consistent

(b) $A^H y \in K^* \Rightarrow \mathrm{Re}(b,y) \geq 0.$

Proof. We show in the following lemma that N(A) + K is closed if and only if AK is closed. Thus under the assumption of Theorem 3.2, consistency and asymptotic consistency coincide and the equivalence of (a) and (b) follows from Theorem 3.1.

Remark. This solvability theorem may be applied to convex sets via their associated cones mentioned in the end of the previous section.

Lemma 3.1 (R.A. Abrams). Let $K \subseteq C^n$. Then AK is closed if and only if N(A) + K is closed.

Proof. Let AK be closed and let $x_k \in K, z_k \in N(A)$ and $y_k = x_k + z_k \to y$. Then $Ay_k = Ax_k \to Ay$. By the assumption $Ay \in AK$, i.e. there exists a vector $x \in K$ such that $Ay = Ax$. Thus $y - x \in N(A)$ and $y = x + y - x \in K + N(A)$.

Conversly, let $N(A) + K$ be closed and let $x_k \in K$ be such that $Ax_k \to b$. Decompose $x_k = y_k + z_k$, where $y_k \in R(A^H)$ and $z_k \in N(A)$. Then $Ax_k = Ay_k$ and $y_k = A^+ A x_k \to A^+ b$, where A^+ is the Moore-Penrose generalized inverse of A, (see section 11).

Thus the sequence $\{y_k\} = \{x_k - z_k\}$ converges to $A^+ b$ and so by the assumption, $A^+ b \in K + N(A)$, say $A^+ b = x - z$, where $x \in K$ and $Az = 0$. To complete the proof it is enough to show that $Ax = b$.

Indeed $Ax = AA^+ b + Az = AA^+ b = AA^+ \lim_{k \to \infty} Ax_k = \lim_{k \to \infty} AA^+ Ax_k = \lim_{k \to \infty} Ax_k = b$.

Corollary 3.1. If K is a polyhedral cone in C^n and $A \in C^{m \times n}$, then $N(A) + K$ is closed.

Proof. Let $K = BR_+^K$. Then $AK = ABR_+^K$ is polyhedral and thus closed.

Theorem 3.2 is the basis of the theory, to be outlined in Chapter 2, of mathematical programming over convex cones which satisfy the closedness condition of the theorem. By Corollary 3.1, these cones include the polyhedral cones. Theorem 3.2 may be derived from similar solvability theorems in general spaces, e.g. Fan [2]. We now mention two of its classical corollaries.

Corollary 3.2 (Farkas [1]). Let $A \in R^{m \times n}$, $b \in R^m$. Then the following are equivalent:

(a) The system

$$Ax = b, \ x \geqq 0$$

is consistent

(b) $$A^T y \geqq 0 \Rightarrow (b, y) \geqq 0.$$

<u>Proof,</u> Choose $K = R_+^n = K^*$ in Theorem 3.2.

<u>Corollary 3.3</u> (Levinson [1]). Let $A \in C^{m \times n}$, $b \in C^m$, $\alpha = (\alpha_i) \in R^n$,

where $0 \leq \alpha_i \leq \frac{\pi}{2}$. Then the system

$$Ax = b, \quad | \arg x_i | \leq \alpha_i, \quad i = 1,2,\ldots,n.$$

is consistent if and only if

$$\arg | (A^H y)_i | + \alpha_i \leq \frac{\pi}{2}, \quad i = 1,2,\ldots,n \Rightarrow \mathrm{Re}(b,y) \leq 0.$$

<u>Proof.</u> Choose $K = T_\alpha$ (and $K^* = T_{\frac{\pi}{2} e - \alpha}$) in Theorem 3.2.

The condition that $N(A) + K$ is closed is in a sense also necessary for the two statements of Theorem 3.2 to be equivalent. This is stated in the following extended version of Theorem 3.2.

<u>Theorem 3.3</u> (Berman [3]). Let $A \in C^{m \times n}$, $S \subset C^m$ and $T \subset C^m$ (S and T sets, not necessarily convex cones). Consider the following statements

(I) The system

   $b - Ax \in T, \quad x \in S$

   is consistent

(II) $y \in T^*$, $A^H y \in S^* \Rightarrow \mathrm{Re}(b,y) \geq 0$.

(These are the statements of Theorem 3.2 when $T = \{0\}$ and S is a closed convex cone).

Then the following are equivalent:

(a) Statements (I) and (II) are equivalent <u>for every</u> $b \in C^n$.

(b) $AS + T$ is a closed convex cone and $AS \cup T \subset AS + T$.

(c) $N[A,I] + S \times T$ is a closed convex cone and $AS \cup T \subset AS + T$.

<u>Proof</u> First notice that $(I) \Longleftrightarrow b \in AS + T$ while $(II) \Longleftrightarrow b \in ((AS)^* \cap T^*)$.

Thus (a) may be rewritten as

(a') $AS + T = ((AS)^* \cap T^*)^*$.

We shall show that $(a') \Longleftrightarrow (b) \Longleftrightarrow (c)$.

$(a') \Rightarrow (b)$. $AS + T$ is a closed convex cone being the dual of a set. Taking duals of the two sides of (a') one gets

$$(AS + T)^* = ((AS)^* \cap T^*)^{**} = (AS)^* \cap T^*.$$

since the intersection of two closed convex cones is a closed convex cone. Thus $(AS + T)^* \subset (AS)^*$ and $(AS + T)^* \subset T^*$, implying $AS \subset AS^{**} \subset (AS + T)^{**} = AS + T$ and $T \subset T^{**} \subset (AS + T)^{**} = AS + T$ so that $AS \cup T \subset AS + T$.

$(b) \Rightarrow (a')$. It is always true that $(AS)^* \cap T^* \subset (AS + T)^*$. Also $AS \cup T \subset AS + T \Rightarrow (AS + T)^* \subset (AS)^* \cap T^*$. Thus $(b) \Rightarrow (AS)^* \cap T^* = (AS + T)^*$ and $(AS)^* \cap T^*)^* = (AS + T)^{**} = AS + T$ since $AS + T$ is a closed convex cone.

$(b) \Longleftrightarrow (c)$. Since the last parts of the conditions are identical one has to show that $AS + T$ is a closed convex cone (in C^m) if and only if $N[A,I] + S \times T$ is a closed convex cone (in $C^{m \times n}$).

This follows from Lemma 3.1 since AS + T = [A,I] (S × T).

Notice that the last part of conditions (b) and (c) holds if $0 \in AS \cap T$ and thus is satisfied if S and T are cones.

We now give a sufficient condition for N(A) + K to be closed, which will be used in the third chapter.

Lemma 3.2 (Berman and Ben-Israel [1]). Let $A \in C^{m \times n}$ and let K be a closed convex cone in C^n. Then N(A) + K is closed if N(A) ∩ K is a subspace of C^n.

Proof. If N(A) ∩ K = {0} the lemma follows from Bourbaki [1], p. 78, Ex. 10. The general case may be reduced to this situation by considering the quotient space C^n/N(A) ∩ K.

Lemma 3.2 and similar results follow from conditions for the sum of two convex sets, and particularly cones, given in Rockafeller [1] Chapter 9, Berman [1] Chapter 3, Fan [2] and Ritter [1]. A finite dimensional corollary of the last two references will be given in the following section.

The consistency of a system which is more general then Ax = b, x ∈ K, is given in the following theorem. The proof is similar to the one of theorem 3.2 and is thus omitted

Theorem 3.4 (Berman [2]). Let $A \in C^{m \times n}$, $b \in C^m$ and let C be an Hermitian positive semi definite matrix of order n and K a closed convex cone in C^n such that N(A) + K is closed. Then the following are equivalent:

(a) The system

(1) $Ax - Cy = b$

(2) $x \in K$

(3) $A^H y \in K*$

(4) $y*Cy \leqq 1$

  is consistent

(b) $A^H z \in K* \Rightarrow Re(b,z) + (z*Cz)^{\frac{1}{2}} \geqq 0.$

Remarks. For $C = 0$, the theorem reduces to theorem 3.2.

2. Choosing $S = T_\alpha$, one gets a solvability theorem of Kaul [1].

The section is concluded with the study of systems of linear inequalities over the interior of cones. Those are of interest in matrix theory, as will be shown in the third chapter.

Theorem 3.5 (Bérman and Ben-Israel [3] [4]). Let $A \in C^{m \times n}$, $b \in C^m$, K a solid convex cone in C^n. Then the following are equivalent:

(a) The system

(3.3) $Ax = b, \; x \in int \; K$

  is consistent

(b) $b \in R(A)$ and $0 \neq A^H y \in K* \Rightarrow Re(b,y) > 0.$

Proof. Let E denote the manifold $E = \{x \mid Ax = b\}$.

If E is empty then both (a) and (b) are false which proves the theorem.

Suppose then, that $E \neq \emptyset$. In this case we show that ($\sim$a), the negation of (a), is equivalent to ($\sim$b). Suppose (a) is not true, so that $E \cap \text{int } K = \emptyset$. Then by Mazur's theorem, Theorem 1.2, there exists a nonzero vector z, such that

$$(3.4) \qquad Ax = b \Rightarrow \text{Re}(x,z) = c \;, \; c \leq 0.$$

$$(3.5) \qquad x \in \text{int } K \Rightarrow \text{Re}(x,z) > 0$$

(3.4) $\Rightarrow z \in R(A^H)$. Say $z = A^H y$ for some $y \neq 0$. Thus $Ax = b \Rightarrow \text{Re}(x,A^H y) \leq 0 \Rightarrow \text{Re }(b,y) \leq 0$. But by (3.5), $0 \neq z = A^H y \in K^* \Rightarrow (\sim b)$.

Assume now that (b) is false. Then there exists a y such that

$$0 \neq A^H y \in K^*, \; \text{Re }(b,y) \leq 0$$

Then

$$x \in E \Rightarrow \text{Re}(Ax,y) \leq 0 \Rightarrow \text{Re}(x,A^H y) \leq 0 \Rightarrow x \notin \text{int } K \Rightarrow (\sim a).$$

A useful corollary of Theorem 3.5 is:

Theorem 3.6 (Berman and Ben-Israel [1]). Let $A \in C^{m \times n}$, and let K_1 and K_2 be solid convex cones in C^m and C^n respectively. Then the following are equivalent:

(a) The system

(3.6) $Ax \in$ int K_1, $x \in$ int K_2

   is consistent

(b) $- y \in K_1^*$, $A^H y \in K_2^* \Rightarrow y = 0$.

Proof. The system (3.6) may be rewritten as:

(3.7) $[A,-I] \binom{x}{z} = 0$, $\binom{x}{z} \in$ int $K_2 \times$ int $K_1 =$ int $(K_2 \times K_1)$ (by Corollary 2.3).

By Theorem 3.5, the system (3.7) is consistent if and only if

(3.8) $0 \in R([A,-I])$

and

(3.9) $0 \neq \begin{pmatrix} A^H \\ -I \end{pmatrix} y \in (K_2 \times K_1)^* \Rightarrow Re(0,y) > 0$.

Now, (3.8) is trivially satisfied, and the conclusion of the
implication (3.9) is impossible. Therefore (3.7) is consistent if and
only if

$$0 \neq [\begin{smallmatrix} A^H \\ -I \end{smallmatrix}] \, y \in (K_2 \times K_1)^*$$

is inconsistent. Using Theorem 2.1(h), this is equivalent to:

$$A^H y \in K_2^*, \ -y \in K_1^* \Rightarrow y = 0.$$

4. Theorems of the alternative

Theorems of the alternative list two systems, exactly one of which is consistent. Six theorems of this type are collected in this section with their corollaries. The first three, given in Ben-Israel [2], are restricted to polyhedral cones. They are stated here without proof.

<u>Theorem 4.1</u> (Ben-Israel [2]). Let

$$A_i \in C^{m \times n_i} (i = 1, \ldots, 4),$$

$$A_1 \neq 0, A_2 \neq 0,$$

T a polyhedral cone in C^m,

S_i polyhedral cones in C^{n_i}, $(i = 1, 2, 3)$,

S_1 pointed and S_2 solid.

Then exactly one of the following two systems is consistent:

$$(I) \sum_{i=1}^{4} A_i x_i \in T, \left\{ \begin{array}{l} 0 \neq x_1 \in S_1, \ x_2 \in S_2 \\ \text{or} \\ x_1 \in S_1, \ x_2 \in \text{int } S_2 \end{array} \right\}, \ x_3 \in S_3$$

$$(II) \ y \in -T^*, \ A_1^H y \in \text{int } S_1^*, \ 0 \neq A_2^H y \in S_2^*, \ A_3^H y \in S_3^*, A_4^H y = 0$$

The real version of this theorem reduces to a theorem of Slater [1], if $T = \{0\}$ and $S_i = R_+^{n_i}$, ($i = i, \ldots, 4$).

Theorem 4.2 (Ben-Israel [2]). Let $T, A_i (i = 1,3,4)$ and $S_i (i = 1,3)$ be as in Theorem 1. Then exactly one of the following systems is consistent:

(I) $\qquad A_1 x_1 + A_3 x_3 + A_4 x_4 \in T, \ 0 \neq x_1 \in S_1, \ x_3 \in S_3$,

(II) $\qquad y \in -T^*, \ A_1^H y \in \text{int } S_1^*, \ A_3^H y \in S_3^*, \ A_4^H y = 0$.

This theorem extends the <u>transposition theorem</u> of Motzkin [1].

Theorem 4.3 (Ben-Israel [2]). Let $T, A_i (i = 2,3,4)$ and $S_i (i = 2,3)$ be as in Theorem 1. Then exactly one of the following systems is consistent:

(I) $\qquad A_2 x_2 + A_3 x_3 + A_4 x_4 \in T, \ x_2 \in \text{int } S_2, \ x_3 \in S_3$

(II) $\qquad y \in -T^*, \ 0 \neq A_2^H y \in S_2^*, \ A_3^H y \in S_3^*, \ A_4^H y = 0$.

This theorem generalizes the transposition theorem of Tucker [1]. Other (complex) corollaries of the above mentioned theorems include results of Mond and Hanson [2], [3].

Let $m = 3$, T the nonpolyhedral convex cone of all vectors in R^3 forming an angle $\leqq \frac{\Pi}{4}$ with $\begin{pmatrix} 1 \\ 0 \\ -1 \end{pmatrix}$, $n_1 = 1$, $A_1 = \begin{pmatrix} 0 \\ 1 \\ 0 \end{pmatrix}$, $S_1 = R_+$, $n_3 = 3$, $A_3 = \begin{pmatrix} 0 & 0 & 0 \\ 0 & 1 & 0 \\ 0 & 0 & 1 \end{pmatrix}$, and S_3 the nonpolyhedral convex cone of all vectors in R^3 forming an angle $\leq \frac{\pi}{4}$ with $\begin{pmatrix} 1 \\ 0 \\ 1 \end{pmatrix}$.

Then neither (I) nor (II) of Theorem 4.2. are consistent.

This example, (Ben-Israel [2]), points out the difficulty of extending the theorems of Motzkin, Tucker and Slater to general, nonpolyhedral, cones. Taking $A_3 = A_4 = 0$ in the Theorems of Motzkin and Tucker gives the transposition theorems of Gordan and Stiemke, respectively. These theorems can be generalized to general cones.

Theorem 4.4 (Berman and Ben-Israel [4]). Let $A \in C^{m \times n}$, and let K be a solid convex cone in R^n. Then exactly one of the following two systems is consistent:

(a) $\qquad\qquad\qquad Ax = 0, \; x \in \text{int } K.$

(b) $\qquad\qquad\qquad 0 \neq A^H y \in K*.$

Proof. The conslusion $\text{Re}(b,y) > 0$ in Theorem 3.5, cannot hold if $b = 0$.

Choosing $K = R_+^n$ in the real version of the theorem reduces it to the result of Stiemke [1].

Theorem 4.5 (Berman and Ben-Israel [1]). Let $A \in C^{m \times n}$. Let K_1 and K_2 be pointed closed convex cones in C^n and C^m respectively. Then exactly one of the following systems in consistent:

(a) $\qquad\qquad\qquad Ax \in K_2, \; 0 \neq x \in K_1$

(b) $\qquad\qquad\qquad A^H y \in \text{int } K_1*, \; -y \in \text{int } K_2*$

Proof. Follows from Theorem 3.6 by changing the roles of K_1, K_2 and their duals.

For $K_1 = R_+^n$ and $K_2 = \{0\}$, the theorem reduces to the result of Gordan [1].

Corollary 4.1. Let K be a pointed cone in C^n and let $A \in C^{mxn}$. Then $N(A) + K$ is closed if $R(A^H) \cap \text{int } K^* \neq \emptyset$.

Proof. $R(A^H) \cap \text{int } K^* \neq \emptyset \Rightarrow N(A) \cap K = \{0\}$ (By Theorem 4.5 with $K_1 = K$ and $K_2 = \{0\}$) $\Rightarrow N(A) + K$ is closed (By Lemma 3.2).

Let S be a convex set in C^n and K a pointed closed convex cone in C^n. A function f defined on S is said to be convex with respect to K if

$$(4.1) \quad \alpha_1 f(x_1) + \alpha_2 f(x_2) - f(\alpha_1 x_1 + \alpha_2 x_2) \in K \quad \text{for every} \quad x_1, x_2 \in S,$$
$$\alpha_1, \alpha_2 \geqq 0, \alpha_1 + \alpha_2 = 1.$$

The chapter is concluded with an example of a theorem of the alternative for convex functions.

Theorem 4.6 (Fan, Glicksberg and Hoffman [1]). Let S be a convex set in C^n and K a pointed closed convex cone in C^m. Let $f: S \to C^m$ be convex with respect to K. Then either

(a) The system

$$(4.2) \quad f(x) \in - \text{int } K$$

is consistent

or

(b) There exists a nonzero vector $y \in K^*$ such that

$$x \in S \Rightarrow \text{Re}(y, f(x)) \geqq 0.$$

Proof. If (4.2) is consistent then (b) cannot hold.

Let B be the convex hull of $f(S)$. Let $\xi \in B$.

Then $\xi = \sum\limits_{i=1}^{r} \alpha_i f(x_i)$, $x_i \in S$, $\sum\limits_{i=1}^{r} \alpha_i = 1$.

By (4.1), $\xi - f(\Sigma \alpha_i x_i) \in K$

Suppose that (4.2) is not consistent. Then
$- f(\Sigma \alpha_i x_i) \notin \text{int } K$ and (by Corollary 2.3.(a)), $\xi \notin - \text{int } K$.

Thus S and $- \text{int } K$ are disjoint convex sets and since $- \text{int } K$ is

the interior of a convex cone there exists, by Corollary 1.3, a

$0 \neq y$ such that

(4.3) $\text{Re}(y, \xi) < 0$ for $\xi \in - \text{int } K$

(4.4) $\text{Re}(y, \xi) \geqq 0$ for $\xi \in S$

$(4.3) \Rightarrow y \in K^*$ and $(4.4) \Rightarrow \text{Re}(y, f(x)) \geqq 0$ if $x \in s$,

which proves (b).

To relate this theorem to the previous results, let S be the whole space and replace f by a linear operator A. Then Theorem 4.6 becomes: Either $- Ax \in$ int K is consistent or there is $0 \neq y \in K*$, such that $\text{Re}(y, Ax) \geq 0$ for every x, i.e. $\text{Re}(A^H y, x) \geq 0$ for every x, i.e. $A^H y = 0$, and this is Corollary 4.1.

MATHEMATICAL PROGRAMMING OVER CONES

The importance of cones in mathematical programming is clear in infinite dimensional problems. e.g., Fan [2], and Guiniard [1].

In the finite dimensional case, the theory of mathematical programming over cones offers a unified approach, suggested by Ben-Israel [1], to the classical real theory and to the theory of complex mathematical programming, that is the theory of programming in complex variables and functions initiated by Levinson [1].

This theory, presented in this chapter, follows from Theorem 3.2 in the same way that its real version follows from Theorem 3.2's real special case, the Lemma of Farkas. With the exception of Section 7, the material of this chapter is based mostly on Abrams [1] and Abrams and Ben-Israel [2].

5. Linear Programming

Let $A \in C^{m \times n}$, $b \in C^m$, $c \in C^n$ and let $S \subset C^n$ and $T \subset C^m$ be closed convex cones.

Consider the primal and dual linear programming problems.

(P)   minimize Re $c^H x$

subject to

(5.1) $Ax - b \in T$, $x \in S$.

(D) maximize Re $b^H y$

subject to

(5.2) $\qquad\qquad\qquad c - A^H y \in S^*, \; y \in T^*.$

A vector $x^0 \in C^n$ is:

(a) a _feasible solution_ of (P) if x^0 satisfies (5.1).

(b) an _optimal solution_ of (P) if x^0 is feasible and

$Re(c^H x^0) = \min \{ Re \; (c^H x); \; x \;\; \text{feasible}\} \equiv$ the optimal value of (P).

The problem (P) is:

(c) _consistent_ if it has feasible solutions.

(d) _unbounded_ if it is consistent, and if it has feasible

solutions $\{x_k; \; k = 1,2,\ldots,\}$ with $Re(c^H x_k) \to - \infty$.

Consistency and boundedness of (D) and feasibility and
optimality of its solutions, are similarly defined.

The _Lagrangian_ of the problems (P) and (D) is

(5.3) $\qquad\qquad L(x,y) = Re\{(c,x) - (y,Ax-b)\}$

$\qquad\qquad\qquad\quad = Re\{(b,y) + (c-A^H y, \; x)\}.$

The point $(x^0, y^0) \in S \times T^*$ is a _saddle_ point of $L(x,y)$ with respect to S × T* if

$$L(x^o, y) \leqq L(x^o, y^o) \leqq L(x, y^o) \text{ for all } x \in S, y \in T*.$$

A duality relation between (P) and (D) and a characterization of there optimal solutions, (if such solutions exist) are given in the following theorem.

__Theorem 5.1__ (Abrams and Ben-Israel [1]). Let S and T in problems (P) and (D) be polyhedral cones. Then

(a) If one of the problems is inconsistent then the other is inconsistent or unbounded.

(b) Let the two problems be consistent, and let x^o be a feasible solution of (P) and y^o be a feasible solution of (D). Then

(5.4) $$\text{Re } c^H x^o \geqq \text{Re } b^H y^o.$$

(c) If both (P) and (D) are consistent, then they have optimal solutions and their optimal values are equal.

(d) Let x^o and y^o be feasible solutions of (P) and (D) respectively. Then x^o and y^o are optimal if and only if

$$\text{Re}\{(Ax^o - b, y^o) + (c - A^H y^o, x^o)\} = 0$$

or equivalently if and only if

$$\text{Re}(Ax^0-b,y^0) = \text{Re}(c-A^Hy^0, x^0) = 0.$$

(e) The vectors $x^0 \in C^n$ and $y^0 \in C^m$ are optimal solutions of (P) and (D) respectively if and only if the point (x^0, y^0) is a saddle point of $L(x,y)$ with respect to $S \times T^*$; in which case

(5.5) $$L(x^0, y^0) = \text{Re } c^Hx^0 = \text{Re } b^Hy^0.$$

Proof. (a). Let (P) be inconsistent. Then by Theorem 3.3 there exists a vector y^1 such that

$$A^Hy^1 \in S^*, \; - y^1 \in T^*, \; \text{Re}(b^Hy^1) < 0.$$

If (D) is consistent, then for any feasible solution y^0 of (D) and any $t \geq 0$, the vector $y^0 - ty^1$ is also a feasible solution of (D), proving, letting $t \to \infty$, that (D) is unbounded.

A similar proof holds if (D) is inconsistent and (P) is consistent.

(b) $\text{Re } c^Hx^0 \geq \text{Re } y^{0H} Ax^0 \geq \text{Re } b^Hy$ by (5.1) and (5.2).

(c) Consider the following system:

$$(5.6)_\alpha \quad \left[\begin{pmatrix} c \\ -b \\ \alpha \end{pmatrix} - \begin{pmatrix} A^H & 0 \\ 0_H & -A_H \\ -b^H & c \end{pmatrix} \begin{pmatrix} y \\ x \end{pmatrix} \right] \in S^* \times T \times iR, \begin{pmatrix} y \\ x \end{pmatrix} \in T^* \times S.$$

The consistency of (P) and (D) and (5.4) implies that the system $(5.6)_{\alpha_0}$ is consistent for some $\alpha_0 \geq 0$. To prove (c) one has to show that $(5.6)_\alpha$ is consistent for $\alpha = 0$. Since S and T are polyhedral, it follows from Theorem 3.3, and from $(iR)^* = R$, that the consistency of $(5.6)_\alpha$ is equivalent to

$$(5.7)_\alpha \quad \begin{pmatrix} u \\ v \\ w \end{pmatrix} \in S \times T^* \times R, \begin{pmatrix} A & 0 & -b \\ 0 & -A^H & -c \end{pmatrix} \begin{pmatrix} u \\ v \\ w \end{pmatrix} \in T \times S^*$$

$$\Rightarrow \quad Re \{(c,u) - (b,v) + (\alpha,w)\} \geq 0.$$

We show that $(5.7)_0$ is true by considering two cases.

i). $w > 0$. Let $x = \frac{u}{w}$, $y = \frac{v}{w}$. Then the left side of the implication $(5.7)_0$ means that x and y are feasible solutions of (P) and (D) respectively and its right side follows from (5.4).

(ii). $w \leq 0$. Assume that $(5.7)_0$ is false, i.e. there exist $\begin{pmatrix} u^o \\ v^o \end{pmatrix} \in S \times T^*$, $w^o \leq 0$ such that

$$\begin{pmatrix} A & 0 & -b \\ 0 & -A^H & C \end{pmatrix} \begin{pmatrix} u^o \\ v^o \\ w^o \end{pmatrix} \in T \times S^* \text{ and } Re (c,u^o) < Re (b, v^o).$$

Then   Re $\{(c,u^o) - (b, v^o) + (a, w^o)\} < 0$ for all $\alpha \geq 0$, and in

particular for α_o, so that $(5.7)_{\alpha_o}$ and $(5.6)_{\alpha_o}$ are false.   Contradiction.

(d).   Follows from (b) and (c).

(e).   Let (x^o, y^o) be a saddle point of $L(x,y)$, with respect

to $S \times T^*$.   For any $x \in S$: $L(x,y^o) = \text{Re} \left\{(b,y^o) + (c - A^H y^o, x)\right\}$

$$\geq L(x^o, y^o) = \text{Re} \left\{(b, y^o) + (c-A^H y^o, x^o)\right\}$$

Therefore:

(5.8) $\text{Re} (c-A^H y^o, x-x^o) \geq 0$ for any $x \in S$.

Substituting $x = 0$ and $x = 2x^o$ in (5.8) shows that

$\text{Re} (a-A^H y^o, x^o) = 0$ so that

$\text{Re}(c-A^H y^o, x) \geq 0$ for all $x \in S$.

that is , $c-A^H y^o \in S^*$, so that y^o $(\in T^*)$ is a feasible solution of (D).

The feasibility of x^o follows similarly from the left side of

the definition of a saddle point.   Substituting $x = y = 0$ in this

definition implies $\text{Re} (c^H x^o) \leq L(x^o, y^o) \leq \text{Re} (b^H y^o)$, which combined

with (5.4) proves (5.5) and the optimality of x^o and y^o.

Conversely. Let x^o and y^o be optimal solutions of (P) and (D) respectively. Then Re $(c^H x^o)$ = Re $(b^H y^o)$ by (c) and (5.5) follows from (d).

For any $x \in S$:

$$L(x, y^o) = \text{Re } \{(b, y^o) + (c - A^H y^o, x)\}$$

$$\geqq \text{Re } (b, y^o), \text{ since } c - A^H y^o \in S*$$

$$= L(x^o, y^o), \text{ by (5.5)}.$$

Similarly, for any $y \in T*$, $L(x^o, y) \leqq L(x^o, y^o)$ and thus (x^o, y^o) is a saddle point of $L(x,y)$ with respect to $S \times T*$.

Remarks. a. Theorem 5.1 is a symmetric and equivalent form of Theorem 4.6 of Ben-Israel [1], where $T = \{0\}$.

b. For $S = R^n_+$ and $T = R^m_+ [\{0\}]$, the problems are standard [canonical] real linear programs and Theorem 5.1 reduces to the classical duality theorem, the complementarity slackness theorem and the classical characterization of optimal solutions via the Lagrangian function.

c. For $S = T_\alpha$ and $T = T_\beta$, Theorem 5.1 becomes a complex duality theorem, Theorem 4.6 of Levinson [1].

d. The polyhedrality of S and T was used in part (c) to assume that

(5.9)
$$\begin{pmatrix} A^H & 0 \\ 0_H & - A_H \\ -b^H & C^H \end{pmatrix} T^* \times S + S^* \times T \times iR$$

is a closed convex cone.

If (5.9) is not closed, then the relation between (P) and (D) is much more complicated e.g. Ben-Israel, Charnes and Kortanek [1] [2].

e. Consider the problem of

maximize Re $c^H x$

subject to

$b - Ax \in S$ and $Ax - a \in T$ where $A \in C^{m \times n}$, $b \in C^m$, $c \in C^n$ and S and T are closed convex cones in C^m.

For $S = T = R^m_+$ this is an interval programming problem. Theorem 5.1 can be applied to study the dual of the generalized problem, e.g. Berman [5].

6. Quadratic Programming

Let $B \in C^{n \times n}$ be a positive semi definite Hermitian matrix, $A \in C^{m \times n}$, $b \in C^m$, $c \in C^n$, and let $S \subset C^n$, $T \subset C^m$ be polyhedral cones.

Consider the pair of convex quadratic problems:

(QP)   minimize $f(x) = Re (1/2\, x^H Bx + c^H x)$

subject to $Ax - b \in T$, $x \in S$,

(QD) maximize $g(y,z) = \text{Re} (-1/2\ y^H By + b^H z)$

subject to $c + By - A^H z \in S^*, z \in T^*$.

Notice that for $B = 0$, these problems reduce to (P) and (D) of the previous section. Define feasibility, consistency and optimality similarly. Abrams and Ben-Israel developed, using a process of linearization similar to that of Dorn in classical quadratic programming, a duality relation between (QP) and (QD):

Theorem 6.1.(Abrams and Ben-Israel [11]). If either of the problems (QP) or (QD) is not consistent then the other one is not consistent or unbounded.

(b) If x is a feasible solution of (QP) and (y,z) is a feasible solution of (QD), then $f(x) \geq g(y,z)$.

(c) If (QP) has an optimal solution x^o, then there exists a vector z^o such that (x^o, z^o) is an optimal solution of (QD) and $f(x^o) = g(x^o, z^o)$.

(d) If (QD) has an optimal solution (y^o, z^o), then there exists a vector x^o, such that $Bx^o = By^o$, which is an optimal solution of (QP) and $f(x^o) = g(y^o, z^o)$.

(e) Let x and (x,z) be feasible solutions of (QP) and (QD), respectively. Then, x and (x,z) are optimal solutions if, and only if,

(6.1) $\text{Re}(x, c+Bx-A^H z) = \text{Re}(z, Ax-b) = 0.$

Proof. (a). If (QP) is inconsistent, then, by Theorem 3.3 there is a vector z^o satisfying $- A^H z^o \in S^*$, $z^o \in T^*$ and $\text{Re } (b^H z^o) > 0$.

If (y,z) is a feasible solution of (QD), then so are $(y, z+tz^o)$ for all $t \geq 0$ and $g(y, z+tz^o) \to \infty$ as $t \to \infty$, so that (QD) is unbounded. In the other direction. If (QD) is inconsistent, that is, if

$$c - (A^H, -B) \begin{pmatrix} 3 \\ y \end{pmatrix} \in S^*, \quad \begin{pmatrix} 3 \\ y \end{pmatrix} \in T^* \times C^n$$

is inconsistent, then, again by Theorem 3.3., there exists a vector x^o such that

$$Ax^o \in T, \quad B^H x^o = Bx^o = 0, x^o \in S \quad \text{and} \quad \text{Re}(x^o, c \cdot) < 0.$$

If x is a feasible solution of (QP), then so are $x + tx^o$, $t \geq 0$, and $f(x + tx^o) = f(x) + t(c, x^o) \to -\infty$ as $t \to \infty$.

(b). Follows from the definition of a polar and from the fact that for any positive semi definite Hermitian $B \in C^{n \times n}$ and $x_1, x_2 \in C^n$.

$$\text{Re } \left\{ \frac{1}{2} x_1^H Bx_1 + \frac{1}{2} x_2^H Bx_2 - x_1^H Bx_2 \right\} \geq 0.$$

(c). Let x^o be an optimal solution of (QP). It is not difficult to check that x^o is also an optimal solution of the primal linear program:

(L.P)   minimize $\mathrm{Re}(-\frac{1}{2} x^{0H}Bx^0 + x^{0H}Bx + c^Hx)$

subject to $Ax - b \in T$, $x \in S$.

The dual of (L.P) is

(L.D)   maximize $\mathrm{Re}(-\frac{1}{2} x^{0H}Bx^0 + b^Hz)$

subject to $c + Bx^0 - A^Hz \in S*$, $z \in T*$.

By Theorem 5.1.there exists an optimal solution z^0 of (L.D) and

$$\mathrm{Re}\,(b^Hz^0) = \mathrm{Re}\,(x^{0H}Bx^0 + \bar{c}^Hx^0).$$

Again, it is easy to check that (x^0, z^0) is a feasible and optimal solution of (Q.D) and that $f(x^0) = g(x^0, z^0)$.

(d)   Rewrite problem (Q.D) in a form of a primal quadratic program, namely

$$\text{minimize } \mathrm{Re}\left[\frac{1}{2}\,(y^H,\ z^H)\begin{pmatrix} B & 0 \\ 0 & 0 \end{pmatrix}\begin{pmatrix} y \\ z \end{pmatrix} - (0,\ b^H)\begin{pmatrix} y \\ z \end{pmatrix}\right]$$

subject to $(B, -A^H)\begin{pmatrix} y \\ z \end{pmatrix} + c \in S*$, $\begin{pmatrix} y \\ z \end{pmatrix} \in C^n \times T*$

and apply part (c).

(e)   Let   x   and(x,z) be feasible solutions.

Then

(6.2) $f(x) - g(x,z) = \text{Re } (x^H Bx + c^H x - b^H z)$

$= \text{Re } (x^H Bx + c^H x - x^H A^H z + x^H A^H z - b^H z)$

$\geqq 0$ with equality if and only if (6.1) holds,

By (b), this is the only case when x and (x,z) are optimal.

Remarks.

(a) For $S = S^* = R^n_+$, $T = T^* = R^m_+$ and real data, Theorem 6.1 gives a classical result of Dorn [1].

(b) For $S = T_\alpha$ and $T = T_\beta$ (with the appropriate dimensions). Theorem 6.1 reduces to a result of Hanson and Mond [1].

(c) A symmetric form of the duality theorem is

Theorem 6.2 (Abrams and Ben-Israel [1]). Let $B \in C^{n \times n}$ and $D \in C^{m \times m}$ be positive semi definite Hermitian matrices, $A \in C^{m \times n}$, $b \in C^n$, and let $S \subset C^n$, $T \subset C^m$ be polyhedral cones, then the following problems are dual in the sense of Theorem 6.1.

(SQP) minimize $F(x,u) = \text{Re}[\frac{1}{2} u^H Du + \frac{1}{2} x^H Bx + b^H x]$

subject to $Du + Ax - b \in T$, $x \in S$

(SQD) maximize $G(y,z) = \text{Re } [-\frac{1}{2} z^H Dz - \frac{1}{2} y^H By + b^H z]$

subject to $c + By - A^H z \in S^*$, $z \in T^*$.

Special cases include the real results of Dorn [2] and Cottle [1], and the complex result of Mond and Hanson [1].

(d) The existence of an optimal solution is guaranteed in some problems, by the following complex version of the Frank-Wolfe [1] theorem.

Theorem 6.3 (McCallum [1]). Let $f(x) = \text{Re}\,[x, (q+Mx)]$ be the real part of a complex quadratic function which is bounded below on a non-empty polyhedral convex set $X \subset C^n$. Then there exists some $x^o \in X$ such that

$$f(x^o) = \min_{x \in X}\ f(x).$$

7. The complementarity problem

The linear complementarity problem is: Given a vector $q \in R^n$ and a matrix $M \in R^{n \times n}$, find vectors x, y which satisfy the conditions

(7.1) $\qquad\qquad x \geqq 0,\ y \geqq 0,\ y = q + Mx$

(7.2) $\qquad\qquad (x,\ y) = 0.$

The importance of the problem lies in the fact that, for special choices of M and q, it contains the problems of solving dual convex quadratic (and thus also linear) programs, the equilibrum point problem of bimatrix games, e.g. Lemke and Hawson [1], and problems in mechanics,e.g. Ingleton [1].

Algorithms for solving the problem are given in Lemke [1] and Cottle and Dantzig [1].

Interesting classes of matrices, of which some will be mentioned in the third chapter,were studied in conjunction with these algorithms and in search of existence theorems.

In particular, it was proved and reproved, Samelson, Thrall and Wesler [1],Cottle [2], Ingleton [1] and Murty [1] that the linear complementarity problem has a unique solution for every vector q if and only if M is a P-matrix, i.e., all its principal matrices are positive, e.g. Gale and Nikaido [1], Fiedler and Ptak [1].

The linear complementarity problem can be extended in various directions. Maier [1] and Cottle [4] study the parametric linear complementarity problem. McCallum [1] considers a complex linear complementarity problem. Cottle and Dantzig [2] replace the matrix M by a vertical block matrix

$$N = \begin{bmatrix} N^1 \\ \vdots \\ N^n \end{bmatrix} \text{where} \quad N^j \in R^{p_j \times n}$$

and (7.1) by

$$x_j \prod_{i=1}^{p_j} y_i^j = 0, \quad (j=1\ldots n)$$

$$\text{wnere} \qquad q = \begin{bmatrix} q^1 \\ \vdots \\ q^n \end{bmatrix} \text{ and } \quad y = \begin{bmatrix} y^1 \\ \vdots \\ y^n \end{bmatrix}$$

are decomposed in conformity with N.

Non-linear complementarity problems, where $q + Mx$ is replaced by $q + f(x)$, f being a function from R^n to R^n, were studied by, among others, Cottle [3], Karamardian [1], [2], Moré [1] and Habetler and Price [1]. The latter considered the problem over convex cones.

For details on the linear and nonlinear problems the reader is referred to a recent exellent review by Lemke [2].

We now return to the linear problem and consider its following version:

Let K be a polyhedral cone in C^n, $q \in C^n$ and $M \in C^{n \times n}$. Find

(7.3) $\qquad z \in K$ and $w \in K*$ such that $w = q + Mz$

and

(7.4) $\qquad\qquad Re\ (z,w) = 0$

In the problem considered by McCallum, $K = T_\alpha$. Rewriting Theorem 3.3 we observe that

Theorem 7.1 The system (7.3) is consistent if and only if

(7.5) $\qquad y \in K,\ -M^H y \in K* \Rightarrow Re\ (q,y) \geqq 0.$

As in the real case, the problem contains the pairs of programs described in the previous sections.

Theorem 7.2 (Berman [4]). Let $B \in C^{n \times n}$ be a positive semi definite Hermitian matrix, $A \in C^{m \times n}$, $b \in C^m$, $c \in C^n$ and let $S \subset C^n$, $T \subset C^m$ be polyhedral cones. Consider the complementarity problem with

$$(7.6) \qquad M = \begin{bmatrix} B & -A^H \\ A & 0 \end{bmatrix}, \quad q = \begin{bmatrix} c \\ b \end{bmatrix}, \quad z = \begin{bmatrix} x \\ y \end{bmatrix}, \quad K = S \times T^*.$$

Then, a solution of this problem solves the pair of problems (Q.P) and (Q.D) (of Section 6) and vice-versa.

The proof is straight forward.

The complementarity problem equivalent to P and D (of Section 5) is the same as the one in Theorem 7.2, with $B = 0$.

A, not necessarily Hermitian, matrix $M \in C^{m \times n}$ is <u>positive semi definite</u> if Re $z^H M z \geq 0$ for every $z \in C^n$. Notice that M, of (7.6), is positive semi definite.

The section is concluded with an existence theorem.

<u>Theorem 7.3</u> (Berman [4]). Let M be a positive semi definite (not necessarily Hermitian) matrix in a complementarity problem which satisfies (7.5). Then the problem $((7.3), (7.4))$ has a solution.

<u>Proof.</u> Consider the related convex quadratic program.

(Q)   Minimize $f(z) = \text{Re} \left(z, q + \frac{1}{2} (M + M^H) z \right)$

       Subject to $z \in K$, $q + Mz \in K^*$

To prove the theorem it suffices to show that (Q) has an optimal solution $\hat{z}$ and that $f(\hat{z}) = 0$. The first part is guaranteed by Theorem 6.3. To show that $f(\hat{z}) = 0$, consider the dual of (Q).

Maximize $g(u,y) = \text{Re} \; (-\frac{1}{2} u^H (M+M^H) \; u - q^H y)$

Subject to

(7.7) $\qquad\qquad (M+M^H) \; u - M^H y + q \in K^*$

(7.8) $\qquad\qquad y \in K$

From (7.7) and (7.8) it follows that

(7.9) $\qquad\qquad g \equiv \text{Re} \; (y^H (M+M^H) \; u - y^H My + q^H y) \geq 0$

Also the positive semi definiteness of M implies:

(7.10) $\qquad\qquad -g(u,y) \geq g$

since

$$-g(u,y) - g = \text{Re} \; (q^H y + \frac{1}{2} u^H (M+M^H) \; u - y^H (M+M) u$$

$$+ \frac{1}{2} y^H (M+M^H) \; y - q^H y)$$

$$= \frac{1}{2} (u^H - y^H) (M+M^H) (u-y) \geq 0.$$

By (7.9) and (7.10), the maximum of $g(u,y)$ is nonpositive and by Theorem 6.1, $f(\hat{z}) = \max g(u,y)$. The constraints of (Q) imply that $f(\hat{z}) \geq 0$ and so $f(\hat{z}) = 0$, which completes the proof.

Remark. In the case $K = T_\alpha$, McCallum [1], showed that the existence theorem holds for a wider class of matrices.

8. Nonlinear Programming

The methods used in the previous sections were algebraic. General, complex programming problems require analytic methods. A Kuhn-Tucker theory for such problems was developed by Abrams and Ben-Israel [3] and Abrams [2]. The reader interested in the theory and in some of its applications is referred to these references and to those mentioned in the introduction to this chapter. Here we cite a sample result.

A function $f\colon C^n \to C$, is <u>analytic</u> in an open domain if in some neighborhood of every point of that domain it may be represented as an absolutely convergent power series about that point in the n complex variables.

A function $g\colon C^n \to C^m$ is analytic if each of its components $g_i\colon C^n \to C$, $i = 1,\ldots,m$, is analytic.

For an analytic function $f\colon C^n \to C$ and a point $z^0 \in C^n$, $\nabla_z f(z^0) \equiv \left((\partial f / \partial z_i) z^0 \right)$, $i = 1,\ldots,n$, denotes the <u>gradient</u> of f at z^0.

For an analytic function $g\colon C^n \to C^m$, we use the notation

$$D_z \, g(z^0) \equiv \left(\frac{\partial g_i}{\partial z_j} (z^0)\right), \, i = 1,\ldots,m, \, j = 1,\ldots,n.$$

Let $K = \bigcap_{k=1}^{p} H_{u_k}$ be a polyhedral cone in C^n, e.g. Theorem 2.4,

and let $z_0 \in K$. Define $K(z_0)$, the cone K at z_0 as the intersection of

those closed half spaces H_{u_k} which contain z_0 in their boundaries.

If $z_0 \in$ int K, then $K(z_0) = C^n$.

Consider the complex nonlinear programming problem

(8.1) Minimize Re $f(z)$, subject to $g(z) \in K$.

Let z^0 be a feasible point of (8.1), $g(z^0) \in K$. Then the

(Kuhn-Tucker) constraint qualification holds at z^0, if every

$z \in C^n$ such that $[D_z \, g(z^0)]z \in K(g(z^0))$ is tangent to a once

differentiable arc $a(\theta)$, beginning at z^0 and leading into

the feasible region, i.e., $a(0) = z^0$, $g(a(\theta)) \in K$ for $0 \leq \theta \leq \varepsilon$

and $a'(0) = tz$ for some $t > 0$, $\varepsilon > 0$.

Theorem 8.1 (Abrams and Ben-Israel [3]). Let K be a polyhedral cone

in C^m. Let $f: C^n \to C$ and $g: C^n \to C^m$ both be analytic in a

neighborhood of a feasible point z^0 at which the constraint

qualification holds. Then a necessary condition for z^0 to be

a local minimum of the problem (8.1) is that there exists a vector

$u \in [K(g(z^0))]^*$ such that

$$\nabla_z \, f(z^0) = [D_z^H \, g(z^0)]u$$

and

$$\text{Re } (g(z^0), u) = 0$$

It is interesting to point out that the generalized Farkas Lemma, Theorem 3.2, is used in the proof of Theorem 8.1 in the same way as the Farkas Lemma, is used in deriving the usual form of the Kuhn-Tucker Theorem. This form for ℓ inequality constraints and $m-\ell$ equality constraints is obtained from the real version of Theorem 8.1, by taking $K = R_+^\ell \times 0^{m-\ell}$, where $0^{m-\ell}$ is the zero element of $R^{m-\ell}$.

Necessary conditions for the problem

(8.2) Minimize Re $f(w^1, w^2)$ subject to $g(w^1, w^2)$ subject to $g(w^1, w^2) \in K$ and

(8.3) $w^2 = \overline{w^1}.$

where $f: C^{2n} \to C$ and $g: C^{2n} \to C^m$ are analytic and K is a polyhedral cone in C^m, are also obtained in Abrams and Ben-Israel [3]. Sufficient conditions for the problem (8.2), (8.3) are given in Abrams [2], under convexity and concavity assumptions (in the sense of Section 4) on f and g. (The only function f of (8.1) which satisfies the convexity assumption is $f(z) = az+b$). These allow duality theory which in turn cover the theorems of Sections 5 and 6, and may be applied to the problem of Section 7.

A duality result for a problem which does not satisfy the assumptions of Abrams [2], is given in Mond [1], where a variance of Theorem 3.4 is used.

Cones in matrix spaces will be studied in the following chapter. The theory of this chapter is applicable to these cones.

CONES IN MATRIX THEORY

Various results in matrix theory may be obtained via the theory of cones by choosing appropriate matrix operators and matrix cones.

The inner product in C^{mxn} which will be used in this chapter is

$$(X, Y) = tr\ X\ Y^H, \text{ (the trace of } XY^H)$$

Let $T(X): C^{mxn} \to C^{pxq}$ be given by

$$T(X) = \sum_{i=1}^{s} A_i\ XB_i$$

Then its adjoint $T^*(Y)$ can easily be shown to be given by

$$T^*(Y) = \sum_{i=1}^{s} A_i^H Y\ B_i^H.$$

Most of the forthcoming results will concern real spaces of matrices. The inner product mentioned above reduces in R^{mxn} to

$$(X, Y) = tr\ X\ Y^T.$$

and in V, the (real) space of Hermitian complex matrices of order n, to

$$(X, Y) = \text{tr } X Y.$$

Several cones of matrices are studied in the next section. The remaining sections of the monograph describe some applications.

9. Cones of matrices

(a) Let K_1 and K_2 be closed convex cones in R^n and R^m, respectively. Denote by $\Pi(K_1, K_2)$, the set of matrices A in R^{mxn} for which $AK_1 \subseteq K_2$. For $K_1 = R_+^n$ and $K_2 = R_+^m$, $\Pi(K_1, K_2)$ is the set of all nonnegative mxn matrices. The set $\Pi(K_1, K_2)$ is a closed convex cone. If K_1 and K_2 are full cones, then so is $\Pi(K_1, K_2)$. If K_1 and K_2 are polyhedral cones, then so is $\Pi(K_1, K_2)$ (e.g. Schneider and Vidyasagar [1]). If $A \in \Pi(K_1, K_2)$, then $A^T \in \Pi(K_2^*, K_1^*)$.

Let
$$P = \left\{ xy^T; x \in K_2, y \in K_1^* \right\}$$

and
$$Q = \left\{ uv^T; u \in K_2^*, v \in K_1 \right\}$$

Theorem 9.1 (Berman, Gaiha [1]).

a. $\Pi(K_1, K_2) = Q^*$

b. $(\Pi(K_1, K_2))^* = \text{cl conv } Q$

c. $(\Pi(K_1, K_2))^* \subseteq \Pi(K_1^*, K_2^*)$.

<u>Proof</u>. a. $Q^* = \left\{A; \ \text{tr } uv^T A^T \gneqq 0 \ \text{ for } \ u \in K_2{}^*, \ v \in K_1\right\}$

$= \left\{A; \ Av \in K_2 \ \text{ if } \ v \in K_1\right\} = \Pi \ (K_1, \ K_2)$.

b. Follows from a. and Corollary 2.1.

c. Replacing K_1 and K_2 with their duals, part a. becomes:

$\Pi(K_1{}^*, \ K_2{}^*) = P^*$. Now c. follows from Theorem 2.1, since $P \subseteq \Pi(K_1, \ K_2)$.

<u>Corollary 9.1</u>. Let K_1 and K_2 be self dual. Then every mxn **matrix** is the difference $A = B-C$, where $B, \ C \in \Pi(K_1, \ K_2)$ and $\text{tr } BC^T = 0$.

<u>Proof</u>. Follows from part c and from Corollary 2.4.

(b) For $m = n$ and $K = K_1 = K_2$, we denote $\underline{\Pi(K)} = \Pi(K_1, \ K_2)$.

For $K = R_+^n$, $\Pi(K)$ is the set of square nonnegative matrices. The Perron-Frobenius theory of nonnegative matrices has been extended by many authors, e.g. Krasnoselskii [1], Krein and Rutman [1], Marek [1] to operators on a Banach space which leave a cone invariant. Here we shall mention special cases of these results, concerning K-nonnegative matrices, i.e. matrices in $\Pi(K)$, where K is assumed to be a full cone. A matrix A is said to have the <u>Perron Property</u> if $\rho(A)$, the spectral radius of A, is an eigenvalue.

<u>Theorem 9.2</u>. (Finite dimensional Krein and Rutman [1], Birkhoff [1]).

Let K be a full cone and let $A \in \Pi(K)$. Then A has the Perron property and there is a non zero vector $x \in K$ such that

$Ax = \rho(A)x$, and a nonzero vector $y \in K^*$ such that $A^T y = \rho(A)y$.

An important corollary of Theorem 9.2 is

Corollary 9.2. (Schneider [1]). Let K be a full cone. Let $T = R-S$ where $S \in \Pi(K)$ and either $R(\text{int } K) \supseteq \text{int } K$ or $R(\text{int } K) \cap \text{int } K = \emptyset$. Then the following statements are equivalent:

1. R is nonsingular, $R^{-1} > 0$ and $\rho(R^{-1}S) < 1$.

2. T is nonsingular and $T^{-1}(\text{int } K) \subseteq \text{int } K$.

3. $Tx \in \text{int } K$, $x \in \text{int } K$ is consistent.

4. $-T^T y \in K^*$, $y \in K^* \Rightarrow y = 0$

This corollary was applied by Schneider to stability and monotonicity theorems. See sections 10 and 11.

Proof. If $R(\text{int } K) \cap \text{int } K = \emptyset$, then none of the statements holds. If $R(\text{int } K) \supseteq \text{int } K$, then 1. $\Rightarrow R^{-1}S \in \Pi(K)$, $T = R(I-R^{-1}S)$, T^{-1} exists and $T^{-1} \in \Pi(K)$. Being nonsingular, T^{-1} maps open sets to open sets so that $T^{-1}(\text{int } K) \subseteq \text{int } K$ which shows 2. Statement 2. $\Rightarrow$ 3., trivially. Statement 4. follows from 3. by Theorem 3.6 and implies 1. by Theorem 9.2.

A converse of Theorem 9.2, and extension for irreducible and positive matrices, were given by Vandergraft [1]. To state them we need the following definitions.

If λ is an eigenvalue of a matrix A, then the degree of λ is the size of the largest diagonal block, in the Jordan canonical form of A, which contains λ.

A matrix $A \in \Pi(K)$ is K-irreducible if it leaves invariant

no proper face of K. A matrix A ε π(K) is **K-positive** if A(K-{0})⊆ int K.
The set of K-positive from the interior of matrices form Π(K), e.g. Barker [1].

Theorem 9.3. If K is a full cone, and A ∈ Π(K), then A has
the Perron property and the degree of ο (A) is no smaller then the
degree of any other eigenvalue having the same modulus.

Furthermore, if A has these two properties, then A ∈ Π(K)
for some full cone K.

Corollary 9.3. If A is a symmetric matrix, then either A or - A
leaves some full cone invariant. Also, every strictly triangular
matrix has an invariant full cone.

Theorem 9.4. A ∈ Π(K) is K irreducible if and only if no
eigenvector of A lies on the boundary of K. If A is K -
irreducible, then

(i). ρ (A) is a simple eigenvalue, and any other eigenvalue
with the same modulus is also simple.

(ii). there is an eigenvector, corresponding to ρ(A), in
int K, and no other eigenvector lies in K.

Furthermore, (i) is sufficient for A to be K-irreducible
with respect to some invariant full cone.

Theorem 9.5. If A is K-positive, then

(i). ρ(A) is a simple eigenvalue, greater than the magnitude
of any other eigenvalue.

(ii). an eigenvector corresponding to ρ (A) lies in int K.

Furthermore, condition (i) is sufficient for A to map some full cone into its interior.

A very interesting Perron-Frobenius theory is developed by Schneider and Vidyasagar [1] for classes of matrices which are cross-positive, strongly cross-positive and strictly cross-positive on K. These are classes which correspond and contain $\Pi(K)$, the K-irreducible matrices and the K-positive matrices, respectively.

Other approaches to a Perron-Frobenius theory for $\Pi(K)$, are due to Barker [1], Pullman [1] and Rheinbolt and Vandergraft [1].

We conclude this subsection with a result concerning polyhedral cones.

Theorem 9.6 (Fiedler and Haynsworth [1]). Let $K = PR_+^h \in R^n$ and $K^* = QR_+^k$. Then an nxn matrix A belongs to $\Pi(K)$ if and only if $Q^T AP \geq 0$.

Proof. Denote by X^j the j-th column of X.
Then $A \in \Pi(K) \Rightarrow AP^j \in K$, (j=1,...,h) $\Rightarrow$
$Q^{i^T} AP^j \geq 0$, (i=1,...,k.j=1,...,h) $\Rightarrow Q^T AP \geq 0$.

Conversely $Q^T AP \geq 0 \Rightarrow AP^j \in K$ (j=1,...,h), since $K^{**} = K$.
Thus $A \in \Pi(K)$.

A similar characterization is given by Fiedler and Haynsworth [1], for cones which they name topheavy, that is cones which are symmetric with respect to an axis.

(c). Let <u>PSD</u> denote the closed convex cone of positive

semi definite matrices in V, the space of Hermitian nxn matrices.

<u>Theorem 9.7</u>. (Berman and Ben-Israel [4], Hall [1]). PSD is se-f dual.

<u>Proof</u>. PSD ⊆ PSD*: This is equivalent to A ∈ PSD, B ∈ PSD ⇒ tr AB ≥ 0.

Let A ∈ PSD and B ∈ PSD. Then AoB ∈ PSD, where AoB is

the Hadamard product of A and B, (e.g. Marcus and Minc [1], p. 121,

Theorem 4.5.2) which implies that

$$\big((AoB)x,x\big) \geq 0 \quad \text{for all} \quad x \in C^n.$$

Let e denote a vector of ones. Then

$$\text{tr } AB = \sum_{i,j} a_{ij} \, b_{ij} = \big((AoB)e, e\big) \geq 0.$$

PSD* ⊆ PSD: Let A ∈ PSD. For any x ∈ C^n, xx^H ∈ PSD and therefore

$0 \leq$ tr $Axx^H = (Ax,x)$ which proves that A ∈ PSD.

A corollary of Theorem 9.7 is that PD, the set of positive

definite matrices in V, is the interior of PSD. The generators of

PSD are the Hermitian positive semi definite matrices of rank 1,

since every matrix in PSD of rank ρ is a sum of ρ Hermitian matrices

of rank 1.

For more on the structure of PSD see Tuassky [3][4]. In another

work Tuassky[5] studies a matrix operator on symmetric matrices A:

$$_{1} A \equiv T A T^{T}$$

and observes that $\iota \in \Pi(\mathrm{PSD})$ and thus, by Theorem 9.1 has the spectral radius as an eigenvalue and a corresponding eigenvector which is a positive semi definite matrix. The general question of characterizing $\Pi(\mathrm{PSD})$ seems to be a very difficult one.

Let A_1 and A_2 be two Hermitian matrices of order n. The pencil, $P(A_1, A_2)$, generated by A_1 and A_2 is the set of real linear combinations of A_1 and A_2:

$$P(A_1, A_2) = \left\{ x_1A_1 + x_2A_2;\ x_1,\ x_2 \in R \right\}$$

The question, when does the pencil $p(A_1, A_2)$ contain a positive definite matrix, was studied by many authors, including Au Yeung [1], [2], Kraljevic [1], [2] and Taussky [4], where it was shown that the existence of a positive definite matrix in the pencil, is equivalent to A_1 and A_2 being simultaneously diagonalizable. The question also fits the framework of Theorem 3.6. Indeed, consistency of the system

$$Tx = x_1A_1 + x_2A_2 \in \text{int PSD},\ x = \binom{x_1}{x_2} \in R^2 = \text{int } R^2$$

means the existence of a required positive definite matrix in the pencil, e.g. Berman and Ben-Israel [2].

(d). An $n \times n$ symmetric matrix A is

(i) copositive if $x \geq 0 \Rightarrow (Ax, x) \geq 0$

(ii) completely positive if there are, say, k nonnegative

vectors a_i (i=1,...,k) such that the form $(Ax,x) = \sum_{i=1}^{k} (a_i, x)^2$ for all $x \in R^n$.

Let B,C,P and S denote the sets of completely positive, copositive,
symmetric nonnegative and symmetric positive semi definite, matrices
of order n, respectively. Then B,C,P and S are closed convex cones.
B and C are dual and

$$B^* = C \supset P + S, \quad P \cap S \supset B = C^*.$$

The completely positive and copositive cones have great
importance in combinatorics and mathematical programming. (e.g. the
complementarity problem), See, Hall [1] and Cottle, Habetler and
Lemke [1], [2].

A real matrix A is said to be copositive with respect to a cone $K(\subset R^n)$
if $(Ax,x) \geq 0$ for all $x \in K$. Haynsworth and Hoffman [1] showed that
A has the Perron property if and only if it is copositive with respect
to some self dual cone.

Matrices which are copositive with respect to a cone K are
also called K positive semi definite. A is called K positive definite
if $0 \neq x \in K \Rightarrow (Ax,x) > 0$, and K almost positive definite if

(i) it is K positive semi definite

and (ii) $x \in$ int $K \Rightarrow (Ax,x) > 0$.

Theorem 9.8. Let $A = BB^T$. Then

B x $\in$ int K is consistent if and only if

A is K positive definite

and $0 \neq Bx \in K$ is consistent if and only if

A is K almost positive definite.

Proof. Follows from the theorems of the alternative 4.4 and 4.5.
For $K = R_+^n$ these are theorems 6.1 and 6.2 of Gaddum [2].

10. Lyapunov type theorems

Stable matrices are matrices whose eigenvalues have negative
real parts. They are characterized by Lyapunov theorem (Lyapunov [1],
Bellman [1]).

Theorem 10.1. Let $A \in C^{n \times n}$. Then the following statements are
equivalent:

(i) The matrix equation $A^H X + XA = - I$ has a positive
definite solution X ,

(ii) A is stable.

Proof. Follows from the real part of Theorem 3.5, applied to V
by taking K = PSD, $Ax \equiv TX = A^H X + XA$ and b = - I. See details in
Berman and Ben-Israel [4].

Relatives of the Lyapunov operator $T(X) = AX + XA^H$ were studied
by many authors. In the most general form this was done by Hill [1].

Part of his results, which follow from the theory of linear inequalities over solid cones, are now discribed: Let T: $V \to V$ be given by

$$(10.1) \qquad T(X) = \sum_{i,j=1}^{s} d_{ij} A_i X A_j^H, \text{ where}$$

$$(10.2) \qquad (d_{ij}) \equiv D = D^H \in C^s$$

and the $n \times n$ complex matrices A_1, A_2,...,A_S are <u>simultaneously triangulable</u> i.e., there is a nonsingular matrix Q such that

$$(10.3) \quad Q^{-1} A_i Q \equiv B_i = \begin{pmatrix} \lambda_1^{(i)} & & \\ 0 & \lambda_2^{(i)} & \\ \vdots & & \ddots \\ 0 & \cdots & 0 & \lambda_n^{(i)} \end{pmatrix}, \qquad (i = 1 \ldots s) \;.$$

The following theorem contains two results of Hill which are combined because their proofs, as consequences of Theorem 3.6, are essentially the same. To ease the reading of the theorem, the second result is denoted in square brackets and primed numbering.

<u>Theorem 10.2</u> (Hill [1]). Let the operator T: $V \to V$ be defined by (10.1) (10.2) and (10.3) and let

$$(10.4) \qquad \phi_k = \sum_{i,j=1}^{s} d_{ij} \lambda_k^{(i)} \lambda_k^{(j)} \qquad (k = 1 \ldots n).$$

Then:

    (a) A sufficient condition for the consistency of

(10.5) $T(X) \in PD, \; X \in PD$,

(10.5') $[T(X) \in PD, \; X \in V]$,

is

(10.6) $\phi_k > 0, \; (k = 1,\ldots,n)$.

(10.6') $[\phi_k \neq 0, \; k = (1\ldots n)]$.

    (b) A necessary condition for the consistency of
(10.5) [(10.5')] is

(10.7) $\phi_n > 0$.

(10.7') $[\phi_n \neq 0]$.

    (c) If $A_1,\ldots,A_s$ are quasi commutative (i.e. each A_i
commutes with $A_j A_k - A_k A_j$, $(i,j,k = 1,\ldots,s)$) then (10.6) [(10.6')]
is also a necessary condition for the consistency of (10.5) [(10.5')]
Proof. (Berman and Ben-Israel [1]). The consistency of (10.5) [(10.5')]
is equivalent, by Theorem 3.6 and PD = int PSD [V = int V], to

(10.8) $-T^*(Y) \in PSD, \; Y \in PSD \Rightarrow Y = 0$.

(10.8') $[-T^*(Y) = 0, Y \in PSD \Rightarrow Y = 0]$

where $T^*(Y) = \sum\limits_{i,j=1}^{s} \overline{d_{ij}} \, A_i^H \, Y A_j.$

With this observation we now prove the statements,

    (a) We show that $\sim (10.8) \Rightarrow \wedge (10.6)$ $[\wedge (10.8') \Rightarrow \sim (10.6')]$

where $\wedge$ denotes negation. Let $0 \neq Y \in PSD$ be such that

$-T^*Y \in PSD[T^{*}Y = 0]$. Then $Z = (z_{ij}) \equiv Q^H Y Q$ satisfies $0 \neq Z \in PSD$,

(e.g. Marcus and Minc [1] p. 84) and

$$G = (g_{ij}) \equiv Q^H T^*(Y)Q = \sum\limits_{i,j=1}^{s} d_{ij} B_i^H Z B_j$$

satisfies $- G \in PSD \ [G = 0].$

    Let k be the first integer for which $z_{kk} > 0$. Then the

first $(k-1)$ rows and columns of Z are zero (since $Z \in PSD$) and

$$g_{kk} = \sum\limits_{i,j=1}^{s} d_{ij} \lambda_k^{(i)} \lambda_k^{(j)} z_{kk} = \phi_k z_{kk} \,.$$

Therefore

$$- G \in PSD \Rightarrow g_{kk} \leq 0 \Rightarrow \phi_k \leq 0 \Rightarrow \sim (10.6)$$

$$[G = 0 \Rightarrow \phi_k = 0 \Rightarrow \sim (10.6')].$$

(b). Let X be a solution of (10.5) [10.5']. Then

$$W \equiv Q^{-1}XQ^{-1H} \in PD[W \in V]$$

and

$$F \equiv Q^{-1}T(X)Q^{-1H} = \sum_{i,j=1}^{n} d_{ij}B_i W_j^H \in PD$$

$$\therefore f_{nn} = \phi_n w_{nn} > 0$$

$$\therefore \phi_n > 0 \ [\phi_n \neq 0].$$

(c) Assuming quasi-commutativity we show

$$\sim (10.7) \Rightarrow \sim (10.8)[\sim (10.7') \Rightarrow \sim (10.8')].$$

$A_1^H,\ldots,A_s^H$ are quasi-commutative, since so are $A_1,\ldots,A_s$. Thus for every $k = 1,\ldots,n$ there exists a common eigenvector u_k such that (Drazin, Dungey and Greenberg [1])

$$A_i^H u_k = \lambda_k^{(i)} u_k \ (i = 1,\ldots s).$$

Now

$$0 \neq u_k u_k^H \in PSD \ (k = 1,\ldots,n)$$

and

$$T^*(u_k u_k^H) = \sum_{i,j=1}^{s} \overline{d_{ij}} \overline{\lambda_k^{(i)}} \lambda_k^{(j)} u_k u_k^H = \phi_k u_k u_k^H.$$

Therefore for any $k = 1, \ldots, n$:

$$\phi_k \leq 0 \Rightarrow - T^*(u_k u_k^H) \in PSD \Rightarrow \sim (10.8)$$

$$[\phi_k = 0 \Rightarrow T^*(u_k u_k^H) = 0 \Rightarrow \sim (10.8')].$$

The Lyapunov operator is a special case of the Hill operator where $D = \begin{pmatrix} 0 & 1 \\ 1 & 0 \end{pmatrix}$, $A_1 = I$ and $A_2 = A$. Other important cases are the Schneider operator, Schneider [1], given by $D = \begin{pmatrix} 1 & 0 \\ 0 & -I_{s-1} \end{pmatrix}$, when I_{s-1} is the identity of order s-1, and its special case where $s = 2$ due to Stein [1],.

In the latter case, Theorem 10.2 characterizes <u>converging</u> matrices, i.e. matrices C such that $C^n \to 0$, (Taussky [2]). <u>Corollary 10.1</u> (Stein [1]). Let $C \in C^{n \times n}$. Then all the eigenvalues of C lie in the interior of the unit circle if and only if there exists a positive definite solution to $X - CXC^H$ is positive definite.

If the matrix D in (10.2) has exactly one positive eigenvalue, then more can be said about the operator T. <u>Theorem 10.3.</u> Let T be as in Theorem 10.2 where D has exactly one positive eigenvalue. Let ϕ_k be defined by (10.4). Then the following are equivalent:

(a)   The system

$$T(X) \in PD, \; X \in PD$$

is consistent

(b) T is nonsingular and

$$T(X) \in PD \Rightarrow X \in PD$$

(c) $\phi_k > 0, \; (k = 1, \ldots, n).$

Proof.   Follows from Corollary 9.2 and a theorem of Carlson (p. 139 in Hill [1]).

In Theorem 10.3 Hill showed the equivalence of (a) and (c). The whole theorem was proved by Schneider [1] for $D = (0 - \overset{1}{} \overset{o}{I_s})$ and by Taussky [1] for the Lyapunov operator, extending Theorem 10.1. This extension shows that if A is stable, then the cone

$$C(A) = \{AX + XA^H; \; X \in PSD\}$$

contains PSD. Not much is known about the structure of $C(A)$, e.g. Taussky [4], Loewy [1].

Theorem 10.3 is not valid for matrices D with more than one positive eigenvalue, even if the matrices $A_1, \ldots, A_s$ are quasi-commutative. This is shown by the following example

$$D = \begin{pmatrix} 1 & 0 \\ 0 & 1 \end{pmatrix}, \ A_1 = \begin{pmatrix} 1 & a \\ 0 & 1 \end{pmatrix}, \ A_2 = \begin{pmatrix} 1 & a \\ 0 & 1 \end{pmatrix} \ .$$

Here A_1 and A_2 commute $(A_2 = A_1^{-1})$, but $T(X)$ satisfies (a) and (c) but not (b) if $a \neq 0$.

Let us return to the Lyapunov theorem. It was generalized by Taussky [1] and Ostrowsky and Schneider [1] as follows:

1. Given a complex matrix A there exists an Hermitian matrix X such that $AX + XA^H$ is positive definite if and only if A has no pure imaginary eigenvalues.

2. If $AX + XA^H$ is positive definite then In A = In X, where In$(A) = (\Pi, \nu, \delta)$, Π is the number of eigenvalues of A with positive real part, ν the number with negative real part and δ the number of pure imaginary eigenvalues.

Part 1 of the generalization was extended by Hill and is the bracketed part of Theorem 10.2. Part 2, was extended by him for some classes of matrices D, but this extension does not follow from the linear inequalities representation of the problem and thus is not given here.

The Lyapunov type theorems were extended to Hilbert spaces, e.g. Cain [1], Datko [1].

The Lyapunov operator also appears in the well known theorem of Bellman and Fan [1] on linear inequalities in Hermitian matrices,

which is the basis of their **theory** of mathematical programming
in Hermitian variables. The proof presented here is based on the
theory of Section 3 and appropriate choice of the cones.

<u>Theorem 10.4</u> (Bellman and Fan [1]). Let $A_{ij} \in C^{n \times n}$ be arbitrary,

let B_i, $C_j \in C^{n \times n}$ be Hermitian (i=1,...,p; j = 1,...,q) and **let**

c be a real number. If there exist positive definite Hermitian

matrices Y_i (i=1,...,p) satisfying

$$(10.9) \qquad \sum_{i=1}^{p} (Y_i A_{ij} + A_{ij}^H Y_i) + C_j = 0, \quad (j=1,...,q),$$

then the following statements are equivalent:

    (a) The system

$$(10.10) \quad \left\{ \begin{array}{c} \sum_{j=1}^{q} (A_{ij} X_j + X_j A_{ij}^H) - B_i \in PSD \quad (i=1,...,p) \\[2em] \mathrm{tr} \sum_{j=1}^{q} C_j X_j \geq c \\[2em] X_j = X_j^H, \quad (j=1,...,q) \end{array} \right.$$

    is consistent.

    (b) For any m Hermitian positive semi definite matrices
D_i and any number $d \geq 0$ the relations

$$(10.11) \qquad \sum_{i=1}^{p} (D_i A_{ij} + A_{ij}^H D_i) + dC_j = 0, \quad (j=1,...,q)$$

imply

$$(10.12) \qquad \text{tr} \sum_{i=1}^{p} D_i B_i + dC \leqq 0.$$

<u>Proof</u>. (Berman and Ben-Israel [1]). For notational convenience, the proof is given in the case $p = q = 1$, where the indices of the matrices are omitted. The proof in the general case in similar. The system (10.10) can be rewritten, as

$$(10.13) \quad \begin{cases} T \begin{pmatrix} x \\ u \\ w \end{pmatrix} \equiv \begin{pmatrix} T_A & -I & 0 \\ T_C & 0 & -I \end{pmatrix} \begin{pmatrix} x \\ u \\ w \end{pmatrix} = \begin{pmatrix} B \\ \frac{2c}{n} & I \end{pmatrix} \\ \\ \begin{pmatrix} x \\ u \\ w \end{pmatrix} \in V \times PSD \times W \equiv S \end{cases}$$

where T_L , the Lyapunov operator, is given by

$$(10.14) \qquad T_L X = LX + XL^H$$

and W is the set of Hermitian matrices with nonnegative trace.

If $N(T) + S$ is closed (for T and S defined by (10.13)) then the consistency of (10.13) is equivalent by Theorem 3.2, to

$$(10.15) \qquad \begin{pmatrix} T_A^* & T_C^* \\ -I & 0 \\ 0 & -I \end{pmatrix} \begin{pmatrix} -D \\ -\frac{d}{2} & I \end{pmatrix} \in S^* = \{0\} \times PSD \times W^*$$

implies

(10.16) $\operatorname{tr}(-D)B + (-d)\ c \geq 0$

Substituting $T^*_L = L^H Y + Y L$ and $W^* = \{t\ I,\quad t \geq 0\}.$ this implication becomes (b).

To complete the proof it therefore suffices to show that $N(T) + S$ is closed for which a sufficient condition is, by Lemma 3.2, that $N(T) \cap S$ is a subspace. The latter assertion follows from the existence of the Hermitian matrix Y, assumed in the theorem. Indeed, this assumption states that the system

(10.17) $T^*_A\ Y = -\ C,\ Y \in \operatorname{int}\ PSD$

is consistent, which by Theorem 3.6 is equivalent to

(10.18) $0 \neq T_A\ X \in PSD \Rightarrow \operatorname{tr}\ CX < 0.$

From (10.13)

$$N(T) = \left\{ \begin{pmatrix} X \\ T_A\ X \\ T_c\ X \end{pmatrix} ;\ X \in V \right\}$$

Thus $N(T) \cap S$ consists of the vectors $\begin{pmatrix} X \\ T_A\ X \\ T_c\ X \end{pmatrix}$ where $X \in V,$

(10.19) $\qquad T_A X \in PSD \quad$ and

$$T_c X \in W$$

For such vectors $T_A X = 0$ since otherwise tr $CX < 0$ by (10.18), contradicting (10.19)

Therefore

$$N(T) \cap S = \left\{ \begin{pmatrix} X \\ 0 \\ T_c X \end{pmatrix} ; \; X \in N(T_A) \right\}$$

is a subspace, completing the proof for the special case $p = q = 1$.

For the general case the only modification needed in this proof is changing T and S to:

$$T = \begin{pmatrix} T_{A_{11}} & T_{A_{12}} & \text{---} & T_{A_{1q}} & -I \\ \vdots & & & & \\ T_{A_{p1}} & T_{A_{p2}} & \text{---} & T_{A_{pq}} & & -I \\ T_{c_1} & T_{c_2} & \text{---} & T_{c_q} & & & -I \end{pmatrix}$$

and

$$S = \underbrace{Vx \;\text{---}\; xV}_{q \text{ times}}, \; x \underbrace{PSDx \;\text{---}\; xPSD}_{P \text{ times}} x \; W.$$

11. Cone monotonicity

Let K be a **full** cone in C^n. The nxn matrix A is K-Monotone if

$$(11.1) \qquad Ax \in K \Rightarrow x \in K.$$

This is clearly equivalent to A being K-inverse positive, that is to say, A is nonsingular (since its null space lies, by (11.1) in K) and $A^{-1} \in \Pi(K)$.

For real matrices and for $K = R_+^n$, this concepts and observation are due to Collatz [1], who called K of monotone kind. Varga [1] uses monotone for of monotone kind.

Monotone matrices have important applications in Applied Mathematics, and in particular in iterative methods of numerical analysis. Some of these applications remain valid when monotonicity is replaced by cone monotonicity and for rectangular matrices. This will be shown in the next section.

Generalization of the monotonicity concept to rectangular matrices (operators between two different spaces) calls for the use of two sets, in the two spaces respectively. Such a generalization is possible by extending the relation (11.1) or by extending the concept of inverse positivity. The first approach was taken by Mangasarian [1] for real matrices and by Ben-Israel [3] and Mond [1] for complex matrices. For example, Ben-Israel [3] replaces (11.1) by $Ax \in K_2 \Rightarrow Bx \in K_1$, and characterizes monotonicity of the pair of matrices $\{A,B\}$ with respect to the pair of closed convex cones $\{K_1, K_2\}$. Schröder [1] studies the more general setting

$$Ax \in C \Rightarrow x \in S$$

where C is a convex set and S is a closed set.

To extend the concept of inverse positivity to singular matrices, we has to replace the inverse of A by A^+, its Moore-Penrose generalized inverse, defined by

$$A^+ y = x \quad \text{if} \quad Ax = y, \ x \in R(A^H)$$

$$A^+ y = 0 \quad \text{if} \quad y \in N(A^H)$$

or equivalently as the unique solution of $AXA = A$, $XAX = X$, AX and XA Hermitian. Note that from the definitions it follows that AA^+ is the orthogonal projection on the range of A, while A^+A is the orthogonal projection on $R(A^H)$.

<u>Theorem 11.1</u>   (Ben-Israel [4]). Let $K_1 \subset C^n$ and $K_2 \subset C^m$ be nonempty sets, and let $A \in C^{m \times n}$.

Then the following statements are equivalent,

(a) $A^+ K_2 \subset K_1$

(b) $Ax \in AA^+ K_2 \Rightarrow A^+ Ax \in K_1$

(c) $Ax \in K_2 + N(A^H) \Rightarrow A^+ Ax \in K_1$

(d) $Ax \in AA^+ K_2$, $x \in R(A^H) \Rightarrow x \in K_1$

(e) $Ax \in K_2 + N(A^H)$, $x \in R(A^H) \Rightarrow x \in K_1$

Proof. (a) $\Rightarrow$ (b). Let $Ax = AA^+ u$ for some $u \in K_2$.

Then $A^+ Ax = A^+ u \in A^+ K_2 \subset K_1$

(b) $\Rightarrow$ (c). Let $Ax = u + v$, $u \in K_2$, $v \in N(A)^H)$.

Then $Ax = AA^+ u \in AA^+ K_2$ and thus $A^+ Ax \in K_1$.

(c) $\Rightarrow$ (a). Let $u \in K_2$, $u = Ax - v$, $v \in N(A^H)$, (i.e.

$Ax \in K_2 + N(A^H)$). Then $A^+ u = A^+ Ax \in K_1$ so that $A^+ K_2 \subset K_1$.

(b) $\Rightarrow$ (d), (c) $\Rightarrow$ (e). If $x \in R(A^H)$, then $x = A^+ Ax$.

(d) $\Rightarrow$ (b). The left part of the implication in (b) may be

written as

$$AA^+ Ax \in AA^+ K_2 .$$

Since $A^+ Ax \in R(A^H)$, the right part of the implication in (b)

follows from (d).

(e) $\Rightarrow$ (c) is proved similarly.

Theorem 11.1 was proved for the nonnegative orthants by Berman
and Plemmons [1]. A related result on row-monotonicity extends in
turn the nonsingular results of Mangasarian and Collatz mentioned
above. Similar characterizations may be obtained for other generalized
inverses, e.g. Berman and Plemmons [3].

From now on, the discussion will be confined again to real matrices.
Monotone matrices with nonpositive off diagonal elements, called
M matrices, are of particular interest because of their importance
in numerical analysis and because their share many properties with
positive definite matrices. For example an M matrix is a P matrix
(see section 7), and its eigenvalues has a positive real part, e.g.
Fan [ ], Fiedler and Ptak [1], [2] and Varga [1]. Equivalently,
M matrices are defined as the difference, $k I-C$, where C is a
nonnegative matrix with spectral radius, $\rho(C)$, and $k > \rho(C)$. A
definition of a rectangular M matrix is given by Plemmons [1].

Let K be a full cone in R^n. Haynsworth [1] defined an
M_K - matrix as a matrix of the form $kI - C$ where $C \in \Pi(K)$ and
$k > \rho(C)$. Barker [1] proposed a more general definition, namely
A is a(K)M matrix if A is K monotone and if it may be split
in the form

(11.2) $A = B - C$, $B \in \Pi(K)$, $B^{-1} \in \Pi(K)$ and $C \in \Pi(K)$.
Notice that this definition extends Haynsworth's, since I is both
K-monotone and K-nonnegative. Also, $B \in \Pi(K)$, $B^{-1} \in \Pi(K) \Rightarrow BK = K$.

Barker calls the splitting (11.2), a completely regular splitting, and the splitting

$$A = B - C, \; B^{-1} \in \Pi(K), \; C \in \Pi(K)$$

a regular splitting. The latter definition was given by Varga for $K = R_+^n$, and will be further discussed in the last section. If $T = R - S$ in Corollary 9.2 is a regular splitting, then R and S satisfy the assumptions of the theorem and it follows that T is K-monotone (and if the splitting is complete, a (K)M-matrix) if and only if $R^{-1}S$ converges . In fact, Schneider suggested Corollary 9.2 to point out a link between M-matrices and Lyapunov type theorems. Two corollaries of Corollary 9.2 are now given. The first is stated for a regular splitting and the second for completely regular splittings.

Theorem 11.2. (Barker [1]). Let $A = B - C$ be a regular splitting with respect to a full cone K. Then the following are equivalent:

(a) $A^{-1} \in \Pi(K)$.

(b) the real parts of the eigenvalues of $B^{-1}A$ are positive.

(c) the real eigenvalues of $B^{-1}A$ are positive.

Proof. (a) $\Rightarrow$ $\varrho \, (B^{-1}C) < 1$. The eigenvalues of $B^{-1}A$ are of the form $1-\lambda$ for λ an eigenvalue of $B^{-1}C$, and $\lambda < 1 \Rightarrow Re(1-\lambda) > 0$. It is clear that (b) $\Rightarrow$ (c). Finally, if the real eigenvalues of

$B^{-1}A$ are positive, then so, in particular, is $1-\rho(B^{-1}C)$. Thus $1 > \rho(B^{-1}C)$ and A is K-monotone.

Theorem 11.2 shows that M_K matrices enjoy the positivity property of the spectrum of M matrices. This is, however, not true for (K)M matrices, as shown by the following example of Barker.

Let $K = \left\{ \left(\begin{smallmatrix} x_1 \\ x_2 \end{smallmatrix} \right); \ 0 \leqq \frac{1}{2} x_1 \leqq x_2 \leqq 2 x_1 \right\}$ and $A = \begin{bmatrix} 1 & 0 \\ 5/2 & -1 \end{bmatrix}$

$A = A^{-1} \in \Pi(K)$, is a(K)M-matrix but its eigenvalue are 1 and - 1.

Theorem 11.3. (Barker [1]). Let K be a full cone. Suppose A = B - C is a completely regular splitting. If for any $x \in K$ there is an $\hat{x}$ in K* such that $(\hat{x}, Ax) > 0$, then A is a (K)M matrix.

Proof. $B^{-1}C \in \Pi(K)$ since the splitting is regular. Let $\rho = \rho(B^{-1}C)$ and let $0 \neq y \in K$ be the corresponding eigenvector. From $B^{-1}Cy = \rho y$ it follows that

$$(11.3) \qquad\qquad (\rho B - C) y = 0.$$

Let $\hat{y} \in K*$ be the vector guaranteed by the hypothesis. Then $0 = (\hat{y}, (\rho B - C) y) = (\rho - 1)(\hat{y}, By) + (\hat{y}, (B-C)y)$ by (11.3) and

$$(11.4) \qquad\qquad (1-\rho)(\hat{y}, By) = (\hat{y}, Ay).$$

Now, $(\hat{y}, Ay) > 0$ by the hypothesis and $(\hat{y}, By) \geq 0$ since the splitting is complete. Thus $1-\rho > 0$ and A is K-monotone.

(K)M-matrices for which $A^{-1} \in$ int $\Pi(K)$ are characterized by

Theorem 11.4 (Barker [1]). Let K be a full cone and let A be a (K)M-matrix with the completely regular splitting B-C. Then $A^{-1}(K-\{0\}) \subset$ int K if and only if $B^{-1}C$ is K-irreducible. For the proof and other extentions of classical results on M-matrices, the reader is referred to Barker's paper.

Fiedler and Ptak [3] define a property of "irreducible monotonicity" for rectangular matrices. These are matrices A such that

(11.4) $Ax \geq 0$ for at least one nonzero $x \geq 0$

such that either A is a column vector or no matrix obtained from A by omitting a column, satisfies (11.4). Irreducible M-matrices are "irreducible monotone". Using an equivalant definition, (see Theorem 11.5 below) the concept of irreducible monotonicity is generalized, Berman and Gaiha [1], as follows.

Let K_1 and K_2 be full cones in R^n and R^m, respectively. Let $A \in R^{m \times n}$ and consider the systems

(i) $Ax \in$ int K_2, $x \in$ int K_1

(ii) $A^T y \in K_1^*$, $0 \neq y \in - K_2^*$

(i_0) $Ax \in K_2$, $0 \neq x \in K_1$

$(ii_0)A^T y \in$ int K_1^*, $y \in -$ int K_2^*

By Theorem 3.6, exactly one of the systems (i) and (ii) is consistent, and by Theorem 4.5, this is true for the systems (i_0) and (ii_0).

The set of matrices for which (i) is consistent, is denoted by $S(K_1, K_2)$. The set of matrices for which (ii) is consistent, is denoted by $S_0(K_1, K_2)$. For square matrices and $K_1 = K_2^*$ it follows from the definitions that

$$A + A^T \text{ is positive definite} \Rightarrow A \in S(K_1, K_2)$$

$$A + A^T \text{ is positive semi definite} \Rightarrow A \in S_0(K_1, K_2).$$

For K_1 and K_2 the nonnegative orthants, $S_0(K_1, K_2)$ is the set of matrices which satisfy (11.4).

Theorem 11.5 Let $A \in S_0(K_1, K_2)$. Then the following are equivalent:

(i) $Ax \in K_2$, $0 \neq x \in K_1 \Rightarrow x \in \text{int } K_1$

(ii) $Ax \in K_2$, $x \neq 0 \Rightarrow x \in \text{int } K_1$ or $-x \in \text{int } K_1$ and $Ax = 0$.

The subset of $S_0(K_1, K_2)$ of the matrices which satisfy these statements is denoted by $M(K_1, K_2)$. Notice the similarity between (i) and (11.1). In the case where K_1 and K_2 are the nonnegative orthants, $M(K_1, K_2)$ reduces to the set of irreducible monotone matrices.

We refer the reader to the paper of Berman and Gaiha for a complete study of $M(K_1, K_2)$ matrices. Here we mention several of their properties.

If $K_1 \subset R^n$, then the rank of matrices in $M(K_1, K_2)$ is either n or $n-1$. In the first case, there exists a $y \in \text{int } K_1$ such that $0 \neq Ay \in K_2$, and $0 \neq Ax \in K_2 \Rightarrow x \in \text{int } K_1$. Thus, if $A \in S_0(K_1, K_2)$ then it has rank n and belongs to $M(K_1, K_2)$ if and only if it has a left inverse B, (BA = I), such that $B(K_2) \subseteq \{0\} \cup$ int K_1.

If m = n and $K_1 = K_2 = K$, then the latter condition becomes
$A^{-1} \in$ int $\Pi(K)$. This and Theorem 11.3 proves:

Theorem 11.6 Let K be a full cone and let A be a (K) M-matrix,
with a completely regular splitting, A = B-C, such that $B^{-1}C$ is
K-irreducible. Then $A \in M(K,K)$.

Finally, the generalized irreducible matrices have the
following multiplicative property; If $A \in M(K_1, K_2)$ and $B \in M(K_0, K_1)$,
where K_0, K_1 and K_2 are full cones in spaces of the appropriate
orders, then $AB \in M(K_0, K_2)$.

Matrices in S(K,K) were called by Vandergraft [2], K-semipositive.
They relations with the K-monotone matrices and with positive definite
matrices are combined in the form of the following theorem, which is
related to Corollary 9.2.

Theorem 11.7 (Vandergraft [2]). Let K be a full cone.

(a) If A is nonsingular, then A is K-semipositive if and
only if so is A^{-1}. Also, A is K-monotone if and only if there
exist a nonsingular $B \in \Pi(K)$ such that I-BA $\in \Pi(K)$.

(b) If I-A $\in \Pi(K)$, then the following statements are
equivalent:

 1. A is K-semipositive,

 2. A is K-monotone,

 3. I-A is convergent,

 4. All eigenvalues of A have positive real parts,

5. All real eigenvalues of A are positive.

(c) If A is symmetric, then A is positive definite if and only if there exists a **full** cone K and an $\alpha > 0$, such that A is K semipositive and $I-\alpha K \in \Pi(K)$.

12. Iterative methods for linear systems

The system of linear equations

$$(12.1) \qquad\qquad Ax = b$$

has a unique solution $(A^{-1}b)$ if and only if A is nonsingular. A **splitting** of a nonsingular matrix A is an expression A = B-C, where B is nonsingular. Associated with every splitting of the matrix A in (12.1) is an iterative method

$$(12.2) \qquad\qquad x^{(m+1)} = B^{-1} Cx^{(m)} + B^{-1}b , m \geq 0$$

that converges to $A^{-1}b$ if and only if $B^{-1} C$ is convergent, i.e. all the eigenvalues of $B^{-1} C$ have modolus less than one. If A is expressed as

$$A = D - L - U$$

where D is a diagonal matrix and L and E are respectively strictly

lower and upper triangular matrices and if the diagonal elements
of A are non zero, then the choices B = D, B = D-L and
B = $\frac{1}{w}$ (D-wL), w ≠ 0, give in (12.2) respectively the point Jacobi,
point Gauss-Seidel, and point successive overrelaxation iterative
methods. See for details, Varga [1].

Recall from Section 11, that Varga [1] defined, A = B-C,
to be a regular splitting if B is monotone and C is nonnegative,
Varga also showed that if A = B-C is a regular splitting and if
A is nonotone, then

$$\rho \ (B^{-1}C) \ = \ \frac{\rho \ (A^{-1}C)}{1+\rho(A^{-1}C)} \ < \ 1$$

and thus, the iterative method (12.2) converges. Ortega and
Rheinboldt [1] showed that if A = B-C is a weak regular splitting,
i.e. $B^{-1}C \geq 0$ and $C \geq 0$ then (12.2) converges if and only if A is
monotone. Notice that a regular splitting is clearly a weak regular
splitting. Finally, Vandergraft [2] proved if K is a full cone,
$B^{-1}C \in \Pi(K)$ and B is K-monotone, then (12.2) converges if and only
A is K-semipositive. Notice that by Corollary 9.2 or Theorem 11.7,
A is also K-monotone. Vandergraft [1] [2] and Rheinboldt and
Vandergraft [1], also derived comparison theorems for the iterative
methods mentioned above, in terms of K-nonnegativity, K-monotonicity
and K-semipositivity.

If the matrix A is singular, and in particular rectangular, then (12.1) may have more than one solution or the system may be inconsistent. In these cases one usually seeks the best least squares approximate solution, that is a vector y of mininum norm that minimizes $|| b - Ax ||$. This y is given by A^+b, where A^+ is the Moore-Penrose generalized inverse of A. See section 11. The concept of a regular splitting was recently extended to singular matrices by Berman and Plemmons [2]. In the rest of the paper we describe their approach.

Let $A \in R^{mxn}$. The splitting A = B-C is called a proper splitting of A if R(A) = R(B) and N(A) = N(B). Notice that if A and B are square and nonsingular, then the usual splitting of A is a proper splitting. From the definition of a proper splitting it is easily seen that $I-B^+C$ is nonsingular and that the iterative method

$$(12.3) \qquad x^{m+1} = B^+Cx^m + B^+b, \quad m \geq 0$$

converges to A^+b for every x^o if and only if $\rho(B^+C) < 1$. Characterizations of this condition are given in the following theorem.

Theorem 12.1 (Berman and Plemmons[2]). Let $K_1 \subseteq R^n$ and $K_2 \subset R^m$ be full cones and let A = B-C be a proper splitting of $A \in R^{mxn}$ where $B^+ \in \Pi(K_2, K_1)$ and $B^+C \in \Pi(K_1)$, (a weak regular proper splitting.)

Then the following statements are equivalent:

(i) $A^+ \in \Pi(K_2, K_1)$,

(ii) $A^+C \in \Pi(K_1)$,

(iii) $\rho(B^+C) = \dfrac{\rho(A^+C)}{1+\rho(A^+C)} < 1.$

Proof. The proof depends heavily on Theorem 9.1 and its structure
is: (ii) $\Rightarrow$ (iii) $\Rightarrow \rho(B^+C) < 1 \Rightarrow$ (ii), (i)$\Longleftrightarrow \rho(B^+C) < 1.$ Here
we show that (i) $\Rightarrow \rho(B^+C) < 1.$

$$S_p = \sum_{j=0}^{p=1} (B^+C)^j \quad \text{for every positive integer p.}$$

Then, using the definition of a proper splitting, $S_p B^+ =$

$[I-(B^+C)^P] A^+.$ By (i), $(B^+C)^P A^+ K_2 \subseteq K_1$ and so $[A^+-S_pB^+] K_2 = (B^+C)^P A^+ K_2$

$\subseteq K_1.$ and for every $\ell \in K_2$, $A^+\ell - S_pB^+\ell \in K_1.$ Let $t = A^+\ell$ and

$s_p = S_{p+1} B^+\ell.$ Then the sequence $\{s_i\}_0^\infty$ is monotone nondecreasing and

bounded by t in the partial order induced by K_1, so that it converges, and

(11.4) $\lim (s_i - s_{i-1}) = \lim (B^+C)^i B^+\ell = 0.$

Now, there exists a $0 \neq y \in K_1$ such that

(11.5) $(B^+C)y = \rho(B^+C) y$

since $B^+C \in \Pi(K_1).$

By (11.5), $y \in R(B^+)$, say $y = B^+x.$ Since K_2 is solid, x
can be decomposed as $x = \ell_1 - \ell_2$ for some $\ell_1, \ell_2 \in K_2.$

Then for each positive integer i,

$$[\phi(B^+C)]^i y = (B^+C)^i y = (B^+C)^i B^+ x$$

$$= (B^+C)^i B^+ \ell_1 - (B^+C)^i B^+ \ell_2$$

Thus by (11.4), $[\rho(B^+C)]^i$ converges to 0, or equivalently, $\rho(B^+C) < 1$.

Recall that matrices which satisfy (i) were characterized in Theorem 11.1. Notice also that the equivalence of (ii) and (iii) does not depend on K_2. This extends the nonsingular case results of Barker [1] and Mangasarian [2]. The nonsingular special cases of the equivalence of (i) and (iii) contain the results of Vandergraft, Ortega and Rheinboldt and Varga, mentioned above.

To conclude we cite a theorem which extends a result of Collatz and Schröder on monotone iterations, e.g. Collatz [1]. This is done by removing the requirement that A be nonsingular and by replacing monotonicity by K-monotonicity.

Theorem 12.2. (Berman and Plemmons [2]). Let K be a full cone in R^n and let $A = B-C$ be a proper splitting of $A \in R^{m \times n}$ such that $B^+C \in \Pi(K)$.

(a) If there exist v^0, w^0 such that $v^1 - v^0 \in K$, $w^0 - v^0 \in K$ and $w^0 - w^1 \in K$ where v^1 and w^1 are computed from

$$v^{m+1} = B^+Cv^m + B^+b$$

and

$$w^{m+1} = B^+Cw^m + B^+b$$

then $\{v^i\}_0^\infty$ is a monotone non decreasing sequence bounded above by A^+b and $\{w^i\}_0^\infty$ is a monotone non increasing sequence bounded below by A^+b, with respect to the partial order induced by K. In this case,

$$A^+b = \text{limit } v^i = \text{limit } w^i$$

(b) If $\rho(B^+C) < 1$, then the existence of u^o and w^o is assured.

The ideas of this section may be useful in the study of nonlinear systems. e.g. Rheinboldt [1].

REFERENCES

ABRAMS, R.A.
1. Nonlinear programming in complex space, Ph. D. dissertation, Northwestern University, 1969.
2. Nonlinear programming in complex space: Sufficient conditions and duality, J. Math. Anal. Appl. 38(1972), 619-632.

ABRAMS, R.A. and BEN-ISRAEL, A.
1. A duality theorem for complex quadratic programming, J. Optimization Th. Appl. 4 (1969), 244-252.
2. Complex mathematical programming, Proceedings of the Third Annual Israel Conference on Operations Research, Tel Aviv, July 1969.
3. Nonlinear programming in complex space: Necessary conditions. SIAM J. Control 9 (1971), 606-620.

AU-YEUNG, Y.H.
1. A theorem on a mapping from a sphere to the circle and the simultaneous diagonalization of two Hermitian matrices, Proc. Am. Math. Soc. 20 (1969), 545-548.
2. Some theorems on simultaneous diagonalization of two Hermitian bilinear functions, Glasnik Mat. 6(26) (1971), 3-8.

BOURBAKI, N.
1. Espaces vectoriels topologiques, chap. 1 et 2, Hermann & Cie, Paris, 1953.

BARKER, G.P.
1. On matrices having an invariant cone, Czech. Math. J. 22 (1972), 49-68.

BELLMAN, R.
1. Introduction to matrix analysis, McGraw-Hill, New York, 1960.

BELLMAN, R. and FAN, K.
1. On systems of linear inequalities in Hermitian matrix variables, pp. 1-11 in Proceedings of Symposia in Pure Mathematics, Volume VII: Convexity (edited by V. Klee), Amer. Math. Soc., Providence, R.I. 1963.

BEN-ISRAEL, A.
1. Linear equations and inequalities on finite dimensional, real or complex, vector spaces: A unified theory, J. Math Anal. Appl. 27 (1969), 367-389.
2. Theorems of the alternative for complex linear inequalities, Israel J. Math. 7 (1969), 129-136.
3. On cone-monotonicity of complex matrices, SIAM Rev. 12 (1970).
4. Cone monotonicity. Lecture in the Gatlinburg V meeting on Numerical Algebra. Los-Alamos, June 1972.

BEN-ISRAEL, A. CHARNES, A. and KORTANEK, K.
    1.   Duality and Asymptotic Solvability Over Cones,
        Bull. Amer. Math. Soc. 74, (1969), 318-324.
    2.   Erratum to Duality and Asymptotic Solvability Over
        Convex Cones, Bull. Amer. Math. Soc. 76, (1970), 426.
    3.   Asymptotic duality over closed convex sets, J. Math.
        Anal. and Appl. 35 (1971),677-690.

BERMAN, A.
    1.   Linear inequalities in matrix theory, Ph. D. dissertation,
        Northwestern University, 1970.
    2.   Linear inequalities over complex cones, Centre de Recherches
        Mathématiques Report No. 87 Can. Math. Bull. (Forthcoming).
    3.   Consistency of linear inequalities over sets, Centre de
        Recherches Mathématiques Report No. 109 Proc. Amer. Math.
        Soc.(Forthcoming).
    4.   Complementarity Problem and duality over convex cones,
        Centre de Recherches Mathématiques Report No. 156. Can.
        Math. Bull. (Forthcoming).
    5.   A note on interval programming. Centre de Recherches
        Mathématiques Report No. 194.

BERMAN, A. and BEN-ISRAEL, A.
    1.   More on linear inequalities with applications to matrix
        theory, J. Math. Anal. Appl. 33 (1971), 482-496.
    2.   A note on pencils of Hermitian or symmetric matrices SIAM
        J. Appl. Math. 21, (1971), 51-54.
    3.   Linear inequalities, Mathematical Programming and matrix
        theory, Math. Prog. 1., (1971), 291-300.
    4.   Linear equations over cones with interior: A solvability
        theorem with applications to matrix theory, Lin. Algebra
        and its Appl. (Forthcoming).

BERMAN, A. and GAIHA, P.
    1.   Generalization of irreducible monotonicity, Lin. Algebra
        and its Appl. 5, (1972), 29-38.

BERMAN, A. and PLEMMONS, R.J.
    1.   Monotonicity and the generalized inverse SIAM J. Appl.
        Math. 22 (1972), 155-161.
    2.   Cones and iterative methods for best least squares solutions
        of linear systems. Centre de Recherches Mathématiques
        Report No. 202.
    3.   Binary relations and inverses of nonnegative matrices.
        Centre de Recherches Mathématiques Report No. 208.

CAIN, B.E.
    1.   An inertia theory for operators on a Hilbert space, (preprint).

COLLATZ, L.
    1.   Functional analysis and numerical mathematics, Academic Press, New York, 1966.

COTTLE, R.W.
    1.   Symmetric dual quadratic programs, Quart. Appl. Math. 21 (1963), 237-243.
    2.   Note on a fundamental theorem in quadratic programming, J. Soc. Indust. Appl. Math. 12 (1964), 663-665.
    3.   Nonlinear programs with positively bounded Jacobians, J. Soc. Indust. Appl. Math. 14 (1966), 147-158.
    4.   Monotone solutions of the parametric linear complementarity problem. Operation Research House, Stanford. Report No. 71-19.

COTTLE, R.W. and DANTZIG, G.B.
    1.   Complementary pivot theory of mathematical programming, Lin. Algebra and its Appl. 1 (1968), 103-125.
    2.   A generalization of the linear complementarity problem, J. of Comb. Th. 8 (1970), 79-90.

COTTLE, R.W. HABETLER, G.J. and LEMKE, C.E.
    1.   Quadratic forms semi-definite over convex cones, Proc. Int. Symp. on Math. Prog. Princeton 1967.
    2.   On classes of copositive matrices, Lin. Algebra and its Appl. 3 (1970), 295-310.

DATKO, R.
    1.   Extending a theorem of A.M. Lyapunov to Hilbert space. J. Math. Anal. Appl. 32 (1970), 610-616.

DAY, M.M.
    1.   Normed Linear Spaces, Academic Press, New York, 1962.

DEUTSCH, F.R. and MASERICK, P.H.
    1.   Applications of the Hahn Banach theorem in approximation theory, SIAM Review 9 (1967), 516-530.

DORN, W.S.
    1.   Duality in quadratic programming, Quart. Appl. Math. 18 (1960), 155-162.
    2.   A symmetric dual theorem for quadratic programs, J. Operations Res. Soc. Japan 2 (1960), 93-97.

DRAZIN, M.P. DUNGEY, J.W. and GRUENBERG, K.W.
    1.   Some theorems on commutative matrices, J. London Math. Soc., 26 (1951), 221-228.

EGGLESTON, H.G.
    1.   Convexity, Cambridge University Press, Cambridge, 1958.

FAN, K.
    1.   Convex sets and their applications, Argonne National
        Laboratory Lecture Notes, Argonne, Illinois, Summer 1969.
    2.   A generalization of the Alaoglu-Bourbaki theorem and
        its applications, Math. Zeitschr. 88 (1965), 48-60.
    3.   Inequalities for M-matrices. Indag. Math. 26 (1964),
        602-610.

FAN, K. GLICKSBERG, I. and HOFFMAN, A.
    1.   Systems of inequalities involving convex functions, Proc.
        Amer. Math. Soc. 8 (1957), 617-622.

FARKAS, J.
    1.   Uber die Theorie der einfachen Ungleichungen,
        J. Reine Angew. Math. 124 (1902), 1-24.

FIEDLER, M. and PTAK, V.
    1.   On matrices with non-positive off-diagonal elements and
        positive principal minors, Czech. Math. J. 12 (1962), 382-400.
        382-400.
    2.   Some results on matrices of class K and their application
        to the convergence of iteration procedures. Czech. Math.
        J. 16 (1966), 260-273.
    3.   Some generalizations of positive definiteness and
        monotonicity, Numer. Math. 9 (1966), 163-172.

FIEDLER, M. and HAYNSWORTH, E.
    1.   Cones which are topheavy with respect to a norm. (Preprint).

FRANK, M. and WOLFE, P.
    1.   An algorithm for quadratic programming, Naval Res.
        Logist. Quart. 3 (1956), 95-110.

GADDUM, J.W.
    1.   A theorem on convex cones with applications to linear
        equalities, Proc. Amer. Math. Soc. 3 (1952),
        957-960.
    2.   Linear inequalities and quadratic forms, Pacific J. 8 (1958),
        411-414.

GALE, D. and NIKAIDO, H.
    1.   The Jacobian matrix and the global univalence of mappings,
        Math. Ann. 159 (1965), 81-93.

GORDAN, P.
    1.   Uber die Auflosungen Linearer Gleichungen mit reelen
        Coefficienten, Math. Ann. 6 (1873), 23-28.

GUINIARD, M.
    1.   Generalized Kuhn-Tucker conditions for mathematical
        programming problems in a Banach space SIAM J.
        Control 7. (1969), 232-241.

HABETLER, G.J. and PRICE, A.L.
    1.   Existence theory for generalized nonlinear complementarity
        problems. J. of Opt. Th. 7 (1971), 223-239.

HALL, M.Jr.
    1.   Combinatorial theory, Blaisdell, Waltham, Mass. 1967.

HANSON, M.A. and MOND, B.
    1.   Quadratic programming in complex space, J. Math. Anal.
        Appl. 20 (1967), 507-514.

HAYNSWORTH, E.
    1.   Abstract 667-135, Notices A.M.S., 1969.

HAYNSWORTH, E. and HOFFMAN, A.J.
    1.   Two remarks on copositive matrices, Lin. Algebra and
        its Appl. 2 (1969), 387-392.

HILL, R.D.
    1.   Inertia theorems for simultaneously triangulable complex
        matrices, Lin. Algebra and its Appl. 2 (1969),
        131-142.

HOUSEHOLDER, A.S.
    1.   The Theory of Matrices in Numerical Analysis. Blaisdell
        Publishing Co. New-York 1964.

INGLETON, A.W.
    1.   A problem in linear inequalities, Proc. London Math.
        Soc. 16 (1966), 519-536.

KAUL, R.N.
    1.   On linear inequalities in complex space. Amer. Math.
        Monthly 7 (1970), 956-960.

KARAMARDIAN, S.
    1.   The nonlinear complementarity problem with applications,
        Part 1, J. of Opt. Th. 4 (1969), 87-98.

    2.   The nonlinear complementarity problem with applications,
        Part 2. J. of Opt. Th. 4 (1969), 167-181.

KLEE, V.
1. Some characterizations of convex polyhedra, Acta Mathematica 102 (1959), 79-107.
2. Maximal separation theorems for convex sets, Trans. Amer. Math. Soc. 134 (1968), 133-147.

KRALJEVIC, H.
1. Simultaneous diagonalization of two symmetric bilinear functionals, Glasnik matematicki 21 (1966), 57-63.
2. Simultaneous diagonalization of two σ-Hermitian forms. Glasnik Mat. 5. (25)(1970), 211-216.

KRASNOSELSKII, M.A.
1. Positive solutions of operator equations, Noordhoff, The Netherlands, 1964.

KREIN, M.G. and RUTMAN, M.A.
1. Linear operators leaving invariant a cone in a Banach space, Uspehi Mat. Nauk (N.S.) 3, (23) (1948), 3-95. (English translation in: Amer. Math. Soc. Translations Ser. 1, Vol. 10, pp 199-325, Providence, R.I., 1956).

LEMKE, C.E.
1. Bimatrix equilibrium points and mathematical programming, Management Sci. 11 (1965), 681-689.
2. Recent results on complementarity problems in Nonlinear Programming, (Eds. J.B. Rosen, O.L. Mangamcerian and K. Ritter). Academic Press, New-York, 1970.

LEMKE, C.E. and HOWSON, J.T.
1. Equilibrium points of bimatrix James J. Soc. Indust. Appl. Math. 12 (1964), 413-423.

LEVINSON, N.
1. Linear Programming in complex space, J. Math. Anal. Appl. 14 (1966), 44-62.

LOEWY, R.
1. The minimal eigenvalue of the Lyapunov transform, Ph. D. dissertation, California Institute of Technology, 1972.

LYAPUNOV, A.
1. Problème Général de la Stabilité du Mouvement, Ann. of Math. Studies 17, Princeton Univ. Press, Princeton, N.J., 1947.

MANGASARIAN, O.L.
    1.   Characterizations of real matrices of monotone kind,
        SIAM Rev. 10 (1968), 439-441.
    2.   A convergent splitting of matrices, Numer. Math. 15
        (1970), 351-353.

MARCUS, M. and MINC, M.
    1.   A survey of matrix theory and matrix inequalities,
        Allyn and Bacon, Boston, Mass. 1964.

MAREK, I.
    1. μ_0-positive operators and some of their applications.
        SIAM J. Appl. Math. 15 (1967), 484-494.

MAIER, G.
    1.   Problem- on parametric linear complementarity problems.
        SIAM Rev. 14 (1972),

MOND, B.
    1.   Nonlinear n*n differentiable programming in complex space,
        pp 385-400, in Nonlinear Programming, Academic Press Inc,
        New-York 1970.
    2.   On the monotonicity of complex matrices, SIAM Rev. 12 (1970),
        577-579.

MOND, B. and HANSON, M.A.
    1.   Symmetric duality for quadratic programming in complex space,
        J. Math. Anal. Appl. 23 (1968), 284-293.
    2.   A complex transposition theorem with applications to complex
        programming, Lin. Algebra and its Appl. 2, (1969),
        49-56.
    3.   Some generalizations and applications of a complex
        transposition theorem, Lin. Algebra and its Appl. 2 (1969),
        401-411.

McCALLUM, C.J. Jr.
    1.   The linear complementarity problem in complex space,
        Ph. D. dissertation. Stanford, 1970.

MORE, J.
    1.   The application of variational inequalities to complementarity
        problems and existence theorems. Computer Science Department,
        Connell University, Report No 71-110.

MOREAU, J.J.
    1.   Décomposition orthogonale d'un espace hilbertien selon
        deux cones mutuellement polaires, Comptes. Rend. 255 (1962),
        233-240.

MOTZKIN, T.S.
    1.   Two consequences of the transposition theorem on linear
        inequalities, Econometrica 19 (1951), 184-185.

MURTY, K.J.
    1.   On the number of solutions to the complementarity problem
        and spanning properties of complementary., Lin. Algebra
        and its Appl. 5 (1972), 65-108.

ORTEGA, J. and RHEINBOLDT, W.C.
    1.   Monotone itarations for nonlinear equations with applications
        to Gaus-Seidel methods. SIAM J. Numer. Anal.4 (1967),
        171-190.

OSTROWSKI, A. and SCHNEIDER, H.
    1.   Some theorems on the inertia of general matrices,
        J. Math. Anal. Appl. 4 (1962), pp 72-84.

PLEMMONS, R.J.
    1.   Monotone splittings of rectangular matrices. Math. of
        Computation (Forthcoming).

PULLMAN, N.J.
    1.   A geometric approach to the theory of nonnegative matrices.
        Lin. Algebra and its Appl. 4 (1971), 297-312.

RHEINBOLDT, W.C.
    1.   The numerical solution of nonlinear algebraic systems.
        Lecture Notes NSF-CBMS Regional conference, University
        of Pittsburgh, July 10-14, 1972.

RHEINBOLDT, W.C. and VANDERGRAFT, J.S.
    1.   A simple approach to the Perron-Frobenius theory for positive
        operators on general partially ordered finite-dimensional
        linear spaces.(Preprint).

RITTER, K.
    1.   Optimization theory in linear spaces, I. Math. Ann. 182
        (1969), 189-206.

ROCKAFELLAR, R.T.
    1.   Convex analysis, Princeton University Press, Princeton, N.J.
        1970.

SAMELSON, H. THRALL, R.M. and WESLER, O.
    1.   A partition theorem for Euclidean n-space. Proc. Amer. Math.
        Soc. 9. (1958). 805-807.

SCHAEFER, H.H.
    1.   Topological vector spaces, Macmillan, New York, 1966.

SCHNEIDER, H.
    1.   Positive operators and an inertia theorem, Numer. Math. 7
        (1965), 11-17.

SCHNEIDER, H. and VIDYASAGAR, M.
    1.   Cross-positive matrices. SIAM J. Numer. Anal. 7 (1970),
        508-519.

SCHRODER, J.
    1.   Duality in linear range-domain implications,pp 321-332 in
        Inequalities III. (Ed. O. Shisha) Academic Press, New-York,
        1972.

SLATER, M.L.
    1.   A note on Motzkin's transposition theorem, Econometrica 19
        (1951), 185-186.

STEIN, P.
    1.   Some general theorems on iterants, J. Research Nat. Bur.
        Standards 48 (1952), 11-17.

STIEMKE, E.
    1.   Uber positive Losungen homogener linearer Gleichungen,
        Math. Ann. 76 (1915), 340-342.

TAUSSKY. O.
    1.   A generalization of a theorem of Lyapunov, J. Soc. Ind.
        Appl. Math. 9 (1961), 640-643.
    2.   Matrices C with $C^n \to 0$, J. of Algebra 1 (1964), 5-10.
    3.   Positive-definite matrices, pp 309-319 in Inequalities
        (edited by O. Shisha), Academic Press, New York, 1967.
    4.   Positive definite matrices and their role in the study of
        the characteristic roots of general matrices, Advances in
        Mathematics 2 (1968), 175-186.
    5.   Automorphs of quadratic forms as positive operators.
        pp 341-346 in Inequalities III (Ed. O Shisha). Academic Press,
        New-York, London 1972.

TUCKER, A.W.
    1.   Theorems of alternatives for pairs of matrices, Bull. Amer.
        Math. Soc. 61 (1955), 135.

VANDERGRAFT, J.S.
    1. Spectral properties of matrices which have invariant cones SIAM J. Appl. Math. 16 (1968), 1208-1222.
    2. Applications of partial orderings to the study of positive definitiness, monotonicity and convergence. SIAM J. Numer. Anal 9. (1972), 97-104.

VARGA, R.
    1. Matrix iterative analysis, Prentice Hall, Englewood Cliffs, N.J., 1962.

ZARANTONELLO, E.H.
    1. Projections on convex sets in Hilbert space and spectral theory. pp 237-423 in contributions to nonlinear functional analysis. Academic Press, New-York and London, 1971.

Glossary of Notations

$C^n[R^n]$ denotes the n-dimensional complex [real] vector space.

$C^{mxn}[R^{mxn}]$ denotes the mxn complex [real] matrices.

R_+^n denotes the nonnegative orthant of R^n.

For $A \in C^{mxn}$:

$\bar{A}$ denotes the conjugate,

A^T denotes the transpose,

A^H or A^* denotes the conjugate transpose,

A^+ denotes the generalized inverse,

$R(A)$ denotes the range,

$N(A)$ denotes the null space,

$\text{tr } A$ denotes the trace,

$\sigma(A)$ denotes the spectrum and

$\rho(A)$ denotes the spectral radius.

For $x \in C$:

$\text{Re } x$ denotes the real part,

$\text{arg } x$ denotes the argument.

The inner product of $x, y \in C^n$ is $(x,y) = x^H y$.

The inner product of $A, B \in C^{mxn}$ is $(A,B) = \text{tr } A^H B$.

For a set $S \subseteq C^n$:

int S denotes the interior,

cl S denotes the closure and $S*\{y; x \in S \Rightarrow \text{Re}(x,y) \geq 0\}$ denotes the dual.

If S is a subspace then $S* = S^+$, the orthogonal complement of S.

A nonempty set $S \subset C^n$ is

(a)   a cone, if $\alpha \geq 0 \Rightarrow \alpha S \subseteq S$.

(b)   a convex cone, if it is a cone and if $S + S \subseteq S$.

(c)   a pointed cone, if it is a cone that does not contain a line.

(d)   solid, if int $S \neq \emptyset$.

(e)   a full cone, if it is a pointed, solid, closed convex cone.

(f)   a polyhedral cone, if $S = BR_+^K$ for some $B \in C^{n \times K}$.

For $\alpha = (\alpha) \in R^n$, $0 \leq \alpha_i \leq \frac{\pi}{2}$, T_α denotes the polyhedral cone in C^n :

$$T_\alpha = \left\{ z_i \mid \arg z_i \mid \leq \alpha_i \right\} .$$

The real space of Hermitian matrices of order n, is denoted by V. The closed convex cone of positive semi-definite matrices in V is denoted by PSD. PD denotes the interior of PSD.

For $K_1 \subseteq R^n$, $K_2 \subseteq R^m$:

$$\Pi(K_1, K_2) = \left\{ A \in R^{m \times n}; AK_1 \subset K_2 \right\} ,$$

$$\Pi(K_1) = \Pi(K_1, K_1) .$$

Lecture Notes in Economics and Mathematical Systems

(Vol. 1–15: Lecture Notes in Operations Research and Mathematical Economics, Vol. 16–59: Lecture Notes in Operations Research and Mathematical Systems)

Vol. 1: H. Bühlmann, H. Loeffel, E. Nievergelt, Einführung in die Theorie und Praxis der Entscheidung bei Unsicherheit. 2. Auflage, IV, 125 Seiten 4°. 1969. DM 16,–

Vol. 2: U. N. Bhat, A Study of the Queueing Systems M/G/1 and GI/M/1. VIII, 78 pages. 4°. 1968. DM 16,–

Vol. 3: A. Strauss, An Introduction to Optimal Control Theory. VI, 153 pages. 4°. 1968. DM 16,–

Vol. 4: Branch and Bound: Eine Einführung. 2., geänderte Auflage. Herausgegeben von F. Weinberg. VII, 174 Seiten. 4°. 1972. DM 18,–

Vol. 5: Hyvärinen, Information Theory for Systems Engineers. VIII, 205 pages. 4°. 1968. DM 16,–

Vol. 6: H. P. Künzi, O. Müller, E. Nievergelt, Einführungskursus in die dynamische Programmierung. IV, 103 Seiten. 4°. 1968. DM 16,–

Vol. 7: W. Popp, Einführung in die Theorie der Lagerhaltung. VI, 173 Seiten. 4°. 1968. DM 16,–

Vol. 8: J. Teghem, J. Loris-Teghem, J. P. Lambotte, Modèles d'Attente M/G/1 et GI/M/1 à Arrivées et Services en Groupes. IV, 53 pages. 4°. 1969. DM 16,–

Vol. 9: E. Schultze, Einführung in die mathematischen Grundlagen der Informationstheorie. VI, 116 Seiten. 4°. 1969. DM 16,–

Vol. 10: D. Hochstädter, Stochastische Lagerhaltungsmodelle. VI, 269 Seiten. 4°. 1969. DM 16,–

Vol. 11/12: Mathematical Systems Theory and Economics. Edited by H. W. Kuhn and G. P. Szegö. VIII, IV, 486 pages. 4°. 1969. DM 34,–

Vol. 13: Heuristische Planungsmethoden. Herausgegeben von F. Weinberg und C. A. Zehnder. II, 93 Seiten. 4°. 1969. DM 16,–

Vol. 14: Computing Methods in Optimization Problems. Edited by A. V. Balakrishnan. V, 191 pages. 4°. 1969. DM 16,–

Vol. 15: Economic Models, Estimation and Risk Programming: Essays in Honor of Gerhard Tintner. Edited by K. A. Fox, G. V. L. Narasimham and J. K. Sengupta. VIII, 461 pages. 4°. 1969. DM 24,–

Vol. 16: H. P. Künzi und W. Oettli, Nichtlineare Optimierung: Neuere Verfahren, Bibliographie. IV, 180 Seiten. 4°. 1969. DM 16,–

Vol. 17: H. Bauer und K. Neumann, Berechnung optimaler Steuerungen, Maximumprinzip und dynamische Optimierung. VIII, 188 Seiten. 4°. 1969. DM 16,–

Vol. 18: M. Wolff, Optimale Instandhaltungspolitiken in einfachen Systemen. V, 143 Seiten. 4°. 1970. DM 16,–

Vol. 19: L. Hyvärinen, Mathematical Modeling for Industrial Processes. VI, 122 pages. 4°. 1970. DM 16,–

Vol. 20: G. Uebe, Optimale Fahrpläne. IX, 161 Seiten. 4°. 1970. DM 16,–

Vol. 21: Th. Liebling, Graphentheorie in Planungs- und Tourenproblemen am Beispiel des städtischen Straßendienstes. IX, 118 Seiten. 4°. 1970. DM 16,–

Vol. 22: W. Eichhorn, Theorie der homogenen Produktionsfunktion. VIII, 119 Seiten. 4°. 1970. DM 16,–

Vol. 23: A. Ghosal, Some Aspects of Queueing and Storage Systems. IV, 93 pages. 4°. 1970. DM 16,–

Vol. 24: Feichtinger, Lernprozesse in stochastischen Automaten. V, 66 Seiten. 4°. 1970. DM 16,–

Vol. 25: R. Henn und O. Opitz, Konsum- und Produktionstheorie. I. II, 124 Seiten. 4°. 1970. DM 16,–

Vol. 26: D. Hochstädter und G. Uebe, Ökonometrische Methoden. XII, 250 Seiten. 4°. 1970. DM 18,–

Vol. 27: I. H. Mufti, Computational Methods in Optimal Control Problems. IV, 45 pages. 4°. 1970. DM 16,–

Vol. 28: Theoretical Approaches to Non-Numerical Problem Solving. Edited by R. B. Banerji and M. D. Mesarovic. VI, 466 pages. 4°. 1970. DM 24,–

Vol. 29: S. E. Elmaghraby, Some Network Models in Management Science. III, 177 pages. 4°. 1970. DM 16,–

Vol. 30: H. Noltemeier, Sensitivitätsanalyse bei diskreten linearen Optimierungsproblemen. VI, 102 Seiten. 4°. 1970. DM 16,–

Vol. 31: M. Kühlmeyer, Die nichtzentrale t-Verteilung. II, 106 Seiten. 4°. 1970. DM 16,–

Vol. 32: F. Bartholomes und G. Hotz, Homomorphismen und Reduktionen linearer Sprachen. XII, 143 Seiten. 4°. 1970. DM 16,–

Vol. 33: K. Hinderer, Foundations of Non-stationary Dynamic Programming with Discrete Time Parameter. VI, 160 pages. 4°. 1970. DM 16,–

Vol. 34: H. Störmer, Semi-Markoff-Prozesse mit endlich vielen Zuständen. Theorie und Anwendungen. VII, 128 Seiten. 4°. 1970. DM 16,–

Vol. 35: F. Ferschl, Markovketten. VI, 168 Seiten. 4°. 1970. DM 16,–

Vol. 36: M. P. J. Magill, On a General Economic Theory of Motion. VI, 95 pages. 4°. 1970. DM 16,–

Vol. 37: H. Müller-Merbach, On Round-Off Errors in Linear Programming. VI, 48 pages. 4°. 1970. DM 16,–

Vol. 38: Statistische Methoden I, herausgegeben von E. Walter. VIII, 338 Seiten. 4°. 1970. DM 22,–

Vol. 39: Statistische Methoden II, herausgegeben von E. Walter. IV, 155 Seiten. 4°. 1970. DM 16,–

Vol. 40: H. Drygas, The Coordinate-Free Approach to Gauss-Markov Estimation. VIII, 113 pages. 4°. 1970. DM 16,–

Vol. 41: U. Ueing, Zwei Lösungsmethoden für nichtkonvexe Programmierungsprobleme. VI, 92 Seiten. 4°. 1971. DM 16,–

Vol. 42: A. V. Balakrishnan, Introduction to Optimization Theory in a Hilbert Space. IV, 153 pages. 4°. 1971. DM 16,–

Vol. 43: J. A. Morales, Bayesian Full Information Structural Analysis. VI, 154 pages. 4°. 1971. DM 16,–

Vol. 44: G. Feichtinger, Stochastische Modelle demographischer Prozesse. XIII, 404 Seiten. 4°. 1971. DM 28,–

Vol. 45: K. Wendler, Hauptaustauschschritte (Principal Pivoting). II, 64 Seiten. 4°. 1971. DM 16,–

Vol. 46: C. Boucher, Leçons sur la théorie des automates mathématiques. VIII, 193 pages. 4°. 1971. DM 18,–

Vol. 47: H. A. Nour Eldin, Optimierung linearer Regelsysteme mit quadratischer Zielfunktion. VIII, 163 Seiten. 4°. 1971. DM 16,–

Vol. 48: M. Constam, Fortran für Anfänger. VI, 143 Seiten. 4°. 1971. DM 16,–

Vol. 49: Ch. Schneeweiß, Regelungstechnische stochastische Optimierungsverfahren. XI, 254 Seiten. 4°. 1971. DM 22,–

Vol. 50: Unternehmensforschung Heute – Übersichtsvorträge der Züricher Tagung von SVOR und DGU, September 1970. Herausgegeben von M. Beckmann. VI, 133 Seiten. 4°. 1971. DM 16,–

Vol. 51: Digitale Simulation. Herausgegeben von K. Bauknecht und W. Nef. IV, 207 Seiten. 4°. 1971. DM 18,–

Vol. 52: Invariant Imbedding. Proceedings of the Summer Workshop on Invariant Imbedding Held at the University of Southern California, June – August 1970. Edited by R. E. Bellman and E. D. Denman. IV, 148 pages. 4°. 1971. DM 16,–

Vol. 53: J. Rosenmüller, Kooperative Spiele und Märkte. IV, 152 Seiten. 4°. 1971. DM 16,–

Vol. 54: C. C. von Weizsäcker, Steady State Capital Theory. III, 102 pages. 4°. 1971. DM 16,–

Vol. 55: P. A. V. B. Swamy, Statistical Inference in Random Coefficient Regression Models. VIII, 209 pages. 4°. 1971. DM 20,–

Vol. 56: Mohamed A. El-Hodiri, Constrained Extrema. Introduction to the Differentiable Case with Economic Applications. III, 130 pages. 4°. 1971. DM 16,–

Vol. 57: E. Freund, Zeitvariable Mehrgrößensysteme. VII, 160 Seiten. 4°. 1971. DM 18,–

Vol. 58: P. B. Hagelschuer, Theorie der linearen Dekomposition. VII, 191 Seiten. 4°. 1971. DM 18,–

Vol. 59: J. A. Hanson, Growth in Open Economics. IV, 127 pages. 4°. 1971. DM 16,–

Vol. 60: H. Hauptmann, Schätz- und Kontrolltheorie in stetigen dynamischen Wirtschaftsmodellen. V, 104 Seiten. 4°. 1971. DM 16,–

Vol. 61: K. H. F. Meyer, Wartesysteme mit variabler Bearbeitungsrate. VII, 314 Seiten. 4°. 1971. DM 24,–

Vol. 62: W. Krelle u. G. Gabisch unter Mitarbeit von J. Burgermeister, Wachstumstheorie. VII, 223 Seiten. 4°. 1972. DM 20,–

Vol. 63: J. Kohlas, Monte Carlo Simulation im Operations Research. VI, 162 Seiten. 4°. 1972. DM 16,–

Vol. 64: P. Gessner u. K. Spremann, Optimierung in Funktionenräumen. IV, 120 Seiten. 4°. 1972. DM 16,–

Vol. 65: W. Everling, Exercises in Computer Systems Analysis. VIII, 184 pages. 4°. 1972. DM 18,–

Vol. 66: F. Bauer, P. Garabedian and D. Korn, Supercritical Wing Sections. V, 211 pages. 4°. 1972. DM 20,–

Vol. 67: I. V. Girsanov, Lectures on Mathematical Theory of Extremum Problems. V, 136 pages. 4°. 1972. DM 16,–

Vol. 68: J. Loeckx, Computability and Decidability. An Introduction for Students of Computer Science. VI, 76 pages. 4°. 1972. DM 16,–

Vol. 69: S. Ashour, Sequencing Theory. V, 133 pages. 4°. 1972.
DM 16,–

Vol. 70: J. P. Brown, The Economic Effects of Floods. Investigations
of a Stochastic Model of Rational Investment Behavior in the Face
of Floods. V, 87 pages. 4°. 1972. DM 16,–

Vol. 71: R. Henn und O. Opitz, Konsum- und Produktionstheorie II.
V, 134 Seiten. 4°. 1972. DM 16,–

Vol. 72: T. P. Bagchi and J. G. C. Templeton, Numerical Methods in
Markov Chains and Bulk Queues. XI, 89 pages. 4°. 1972. DM 16,–

Vol. 73: H. Kiendl, Suboptimale Regler mit abschnittweise linearer
Struktur. VI, 146 Seiten. 4°. 1972. DM 16,–

Vol. 74: F. Pokropp, Aggregation von Produktionsfunktionen. VI, 107
Seiten. 4°. 1972. DM 16,–

Vol. 75: GI-Gesellschaft für Informatik e. V. Bericht Nr. 3. 1. Fach-
tagung über Programmiersprachen · München, 9–11. März 1971.
Herausgegeben im Auftrag der Gesellschaft für Informatik von H.
Langmaack und M. Paul. VII, 280 Seiten. 4°. 1972. DM 24,–

Vol. 76: G. Fandel, Optimale Entscheidung bei mehrfacher Zielset-
zung. 121 Seiten. 4°. 1972. DM 16,–

Vol. 77: A. Auslender, Problemes de Minimax via l'Analyse Convexe
et les Inégalités Variationnelles: Théorie et Algorithmes. VII, 132
pages. 4°. 1972. DM 16,–

Vol. 78 : GI-Gesellschaft für Informatik e.V. 2. Jahrestagung, Karlsruhe,
2.–4. Oktober 1972. Herausgegeben im Auftrag der Gesellschaft für
Informatik von P. Deussen. XI, 576 Seiten. 4°. 1973. DM 36,–

Vol. 79 : A. Berman, Cones, Matrices and Mathematical Programming.
V, 96 pages. 4°. 1973. DM 16,–